BIRDS UP CLOSE

BIRDS UP CLOSE

AN ENGINEER EXPLORES THEIR HIDDEN WONDERS

LORNA GIBSON

THE MIT PRESS
CAMBRIDGE, MASSACHUSETTS
LONDON, ENGLAND

The MIT Press
Massachusetts Institute of Technology
77 Massachusetts Avenue
Cambridge, MA 02139
mitpress.mit.edu

The MIT Press's publishing mission benefits from the generosity of our donors, including Cynthia and John Reed.

The MIT Press would like to thank the anonymous peer reviewers who provided comments on drafts of this book. The generous work of academic experts is essential for establishing the authority and quality of our publications. We acknowledge with gratitude the contributions of these otherwise uncredited readers.

This book was set in New Frank and Arnhem Pro by MIT Press. Printed and bound in the United States of America.

Library of Congress Cataloging-in-Publication Data
is available.

ISBN: 978-0-262-04989-4

10 9 8 7 6 5 4 3 2 1

EU Authorised Representative: Easy Access System Europe, Mustamäe tee 50, 10621 Tallinn, Estonia | Email: gpsr.requests@easproject.com

For Jeannie, who delighted in my nature and bird habit
and who encouraged me to write the bird book: love always.

CONTENTS

PREFACE

I'm holding an ivory-billed woodpecker in my hands, a mounted specimen in the ornithology collection of Harvard University's Museum of Comparative Zoology, talking about why woodpeckers peck. And standing next to me is a three-foot-tall section of a bald cypress tree in which an ivory-billed woodpecker has pecked out a nearly two-foot-deep cavity nest.

The woodpecker is somewhat tattered but still magnificent, with a red crest, white edging on its wings, and, of course, an ivory bill. I'm delighted because I love birds (an ivory-billed woodpecker!) and natural history, but at the same time I can't quite believe what I'm doing. I'm a professor of materials science and engineering at MIT, and I never thought I'd be holding an ivory-billed woodpecker, even a museum specimen. I'm here with colleagues from MIT's online education department, making a video about how woodpeckers avoid brain injury when they peck, a project that combines my love of birds, engineering, and teaching.

Birds are amazing.

Consider their feathers, which do so much with color (figure 0.1). They shimmer on the neck of a ruby-throated hummingbird. They glow a brilliant orange on the breast of a Baltimore oriole. On the body of the eastern screech-owl, they match the color and texture of tree bark, creating almost perfect camouflage.

ruby-throated hummingbird

Baltimore oriole

eastern screech-owl

FIGURE 0.1

Feathers provide color and camouflage. [All Alamy.]

And they do so much more. They define the shape of the wing, making flight possible. They insulate against the cold and repel water, allowing eiders, famous for their down, to breed in the Arctic and to winter off the coasts of Newfoundland, Nova Scotia, and New England. They even control sound. The ruff feathers on the face of a barn owl gather sound and direct it to the owl's ears, enabling it to detect prey entirely by hearing, in total darkness—and the feathers on its wings suppress the sound of its motion through the air, allowing it to swoop down on prey virtually silently.

Think about how light birds are, too—an obvious necessity for flight. A house sparrow is about the same length as an eastern chipmunk without its tail, yet it weighs only a third of what a chipmunk does. Bird bones, in particular, are remarkably light. I teach engineering students, and when it comes time to talk about efficient, lightweight structures, I introduce them to bird skulls (figure 0.2). To emphasize just how lightweight they are, I pass a skull from a red-tailed hawk around the class, and when the students hold it in their hands and feel how little it weighs, they gasp in amazement.

Then there are bird bills: the Swiss Army knives of the bird world, used for everything from feeding to preening to nest-building. Bird bills come in diverse sizes and shapes, and are matched to what each species eats. They range from the small, needle-like bills of insect-eating swallows to the dagger-like bills of fish-catching herons to the curved bill that the long-billed curlew uses to probe into sand and mud to capture invertebrates (figure 0.3). The bills of sanderlings can even detect worms and mollusks in wet sand

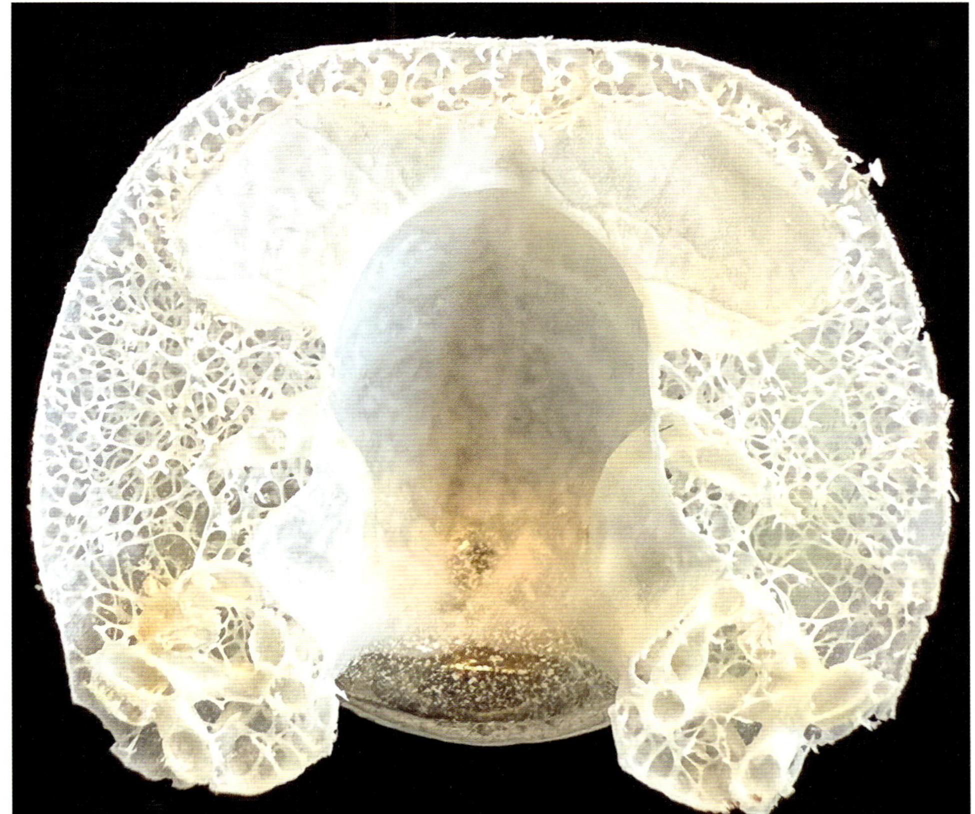

FIGURE 0.2

Cross section of a pigeon skull, showing its lightweight, foam-like interior. [Concord Field Station, Harvard University, specimen CFS3907.]

without touching them. Some birds have spectacular tongues inside their bills: woodpeckers have long tongues to capture insects as they reach far into the holes they drill in trees, and hummingbirds have specially shaped tongues for trapping nectar.

Bird eggs, of course, are marvels of nature. Their shells are at once strong enough to protect the embryo *and* weak enough for chicks to break through when they're ready to hatch. That's no small feat. Eggs come in different shapes: owl eggs are nearly spherical, while those of murres are elongated and more pointed at one end than the other (figure 0.4a, b). Eggs also exhibit remarkable variety in their coloring: from the plain white of woodpecker eggs, hidden in cavity nests, to the speckled eggs of shorebirds, nearly perfectly camouflaged against the sand and pebbles of the open beaches where they are laid (figure 0.4c, d).

FIGURE 0.3

Bird bills come in a remarkable variety of shapes and sizes. [All Alamy.]

barn swallow

great blue heron

long-billed curlew

sanderling

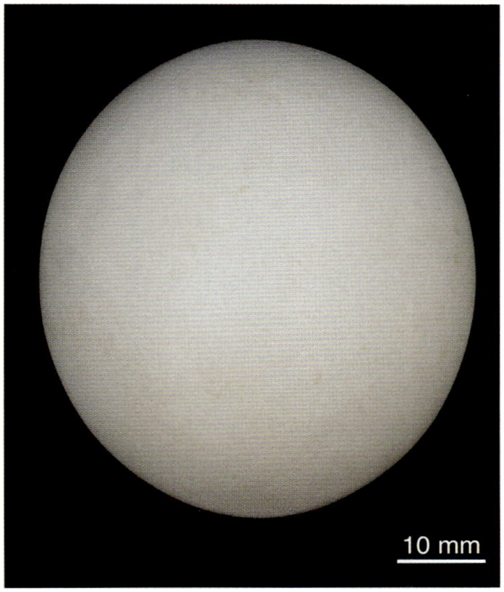

(a) great horned owl egg

(b) common murre egg

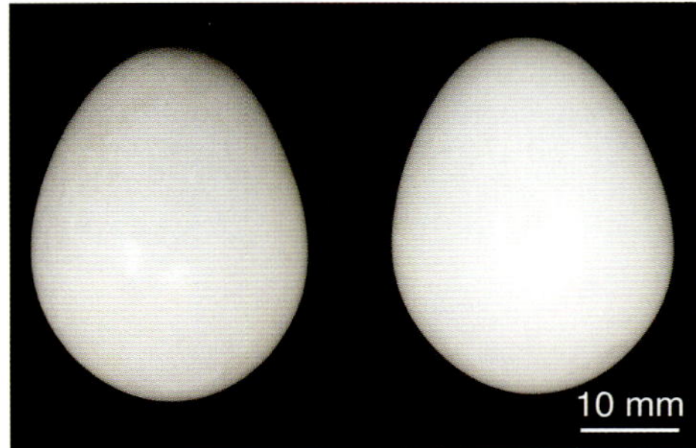

(c) ivory-billed woodpecker eggs

(d) little ringed plover eggs camouflaged on pebbles

FIGURE 0.4

The shapes of bird eggs vary from nearly spherical to elongated and asymmetric. Eggs that are hidden in tree cavities, such as those of owls and woodpeckers, are white, while those exposed on a beach are speckled to blend into the background. [(a, b, c) Museum of Comparative Zoology, Harvard University, specimens 354349, 352849, 361680, © President and Fellows of Harvard; (d) Alamy.]

Beyond all of that, birds fly! What else is there to say? Bird flight is one of the wonders of the natural world. Some, like mallards, flap their wings furiously, while others, like broad-winged hawks, soar and glide with barely any movement. Hummingbirds hover, dart, and even fly backward. Peregrine falcons dive at up to 200 miles per hour (320 kilometers per hour). And great flocks of tens of thousands of starlings dance across the sky together in fantastic murmurations.

Feathers, bones, bills, eggs, flight: how do birds work? That's the question I set out to answer in this book—not just as an avid birder with an interest in natural history but also as an expert in materials science and engineering. (I taught at MIT for nearly 40 years.) Birding is largely a visual endeavor, and so is materials science and engineering. In the same way that birders use binoculars to look closely at the details of a bird's outer appearance to distinguish one species from another, materials scientists and engineers use microscopes to look closely at the internal structure of materials to understand how the structure of a material affects its properties. In this book, we'll be looking at birds more closely, through the eyes of an engineer, sometimes at the microscopic scale, sometimes at a more macroscopic scale, to explain how they work. Chapter by chapter, drawing on both my professional knowledge and my personal experience, I'll explore the hidden microscopic structures and engineering principles behind the astonishing phenomena of birds. Feathers, bones, bills, eggs, and the magic of flight—I'll take stock of them all. With any luck, you'll react much as my MIT colleagues did while I was working on this book. Whenever I stopped them in the hallway to tell them the latest thing I'd learned about how birds work, I always began by saying, "You're not going to believe this," and, after I told them my latest news, we always agreed: birds really *are* amazing. What I've learned about birds in researching this book has amazed me, personally and professionally. But first, let me say just a little about myself and how I came to write this book.

My parents were born and raised in Ashington, a small, gritty coal-mining town north of Newcastle-upon-Tyne, in England. Life there was hard: my father's family were coal miners, working long hours in dangerous conditions. My mother's father managed the Miner's Welfare Institute, which housed facilities to improve the lives of the miners. These included a library, reading room, lecture hall, games room, woodworking shop, and darkroom for photography.

My mother often spoke with great fondness of the first family vacation she took with her parents and sister, bicycling the 30 miles from Ashington to Hepple, a tiny village in the Cheviot Hills, near the border with Scotland. The Cheviots are moorland, covered in heather; between midsummer and autumn, when the heather blooms as far as the eye can see, the hills turn purple. From then on, after that first vacation, she always loved being out in nature, rambling.

After finishing high school, my father went to work in the mine as an apprentice electrician, earning enough money to buy a secondhand Triumph motorbike. When he and my mother were courting (as they used to say), on their days off the two of them would hop onto the Triumph and head up to Rothbury or Wooler in the Cheviots to go fly-fishing and walking. Whenever my parents talked about those times, heading up into the hills on the Triumph, walking among the heather, they got this dreamy, nostalgic look about them; in fact, all they really needed to say was "the Triumph," and in their minds they were there again. Whenever I see a Triumph motorbike, I think of them and their love of walking in the hills, being out in nature.

During his time in the coal mine, my father had a friend, Bob Crook, who told him that the miner's union had a scholarship to send miners to university, based on the results of a competitive national exam. He studied up for the exam at night school in Newcastle, won the scholarship, and became a mechanical engineer. After World War II, my parents emigrated to Canada, eventually ending up in Niagara Falls, Ontario, where I grew up.

Niagara Falls is famously home to many tourist traps around the falls and, less famously, to many large, impressive feats of engineering: a 50-story-tall observation tower near the falls; bridges high up over the gorge and river; and the canals, tunnels, and reservoirs that feed water from above the falls to the hydroelectric power stations below (figure 0.5). Growing up, I wondered how all these structures *worked*. How did the towers and bridges stand up? How did the canals, tunnels, and reservoir manage the huge flow of water to the power station?

My father worked at a company that specialized in the design of hydroelectric power stations. (Canada gets so much of its electricity from hydroelectric power—nearly 60%—that electricity is simply referred to as "hydro." As in, "I paid the hydro bill.")[1] One of my father's responsibilities was running the company's lab, where they made models of the landscape around the power stations that they were designing to get a better understanding of the flow of water before and after a power station was built. The models were huge, often occupying the entire

FIGURE 0.5

The hydroelectric power stations at Niagara Falls, Ontario (left) and New York (right) with their reservoirs above them. [Alamy.]

warehouse-sized lab building; even now, I distinctly recall the sense of awe I had as a small child looking at them.

Our family often took Sunday afternoon walks along the Niagara River, or along the Niagara Escarpment. I especially loved going down the Niagara Glen, climbing down the metal staircase built into the cliff face of the gorge below the falls, hiking along the more rugged trails through old growth forest and past huge boulders; it always seemed like a magical, hidden place, overlooked by most tourists. On our walks, my mother taught my two brothers and me how to identify the common birds, trees, and flowers in our area. In the winter, we always had a bird feeder in the backyard and got the usual customers: sparrows, chickadees, juncos, jays, cardinals, and the occasional woodpecker. In the fall, we were entertained by the unsteady behavior of somewhat inebriated blue jays feeding on fermenting berries on our mountain ash tree. I, too, came to love country walks and nature (figure 0.6).

One summer while I was in high school, on a family trip back to Ashington, we took a boat out to the Farne Islands, a nature reserve just off the coast, home to breeding colonies of thousands of puffins, Arctic terns, guillemots, and kittiwakes. I remember the excitement of seeing the enormously cute puffins and being dive-bombed by Arctic terns protecting their chicks. Shortly after this trip, one of my older brothers became an avid birder, and I started paying more attention to birds at home.

At university, in Toronto, still interested in how those engineering structures in Niagara Falls worked, I pursued an undergraduate degree in civil engineering, mostly to learn more about mechanics: how forces are transmitted through materials and structures, how much force it takes to break something, how fluids flow. And to have a break from engineering, I would take the ferry out to the Toronto Islands, a local park with miles of trails, to walk.

In graduate school, in Cambridge, England, I became more interested in materials. Materials scientists and engineers are interested in how the processing of a material gives rise to a particular microscopic structure, and then how that microscopic structure affects the material's properties and its performance in engineering components. Carbon-fiber composites, for instance, can be produced either with long parallel fibers or short randomly oriented ones; the different arrangements of the fibers (or microstructures of the composites) give rise to different properties that can be tailored for applications ranging from sporting equipment to aerospace components. Looking at the structure of materials in a microscope, and looking for something new or different, is a big part of materials science. It's a little like birders looking for that one Iceland gull among hundreds of herring gulls.

FIGURE 0.6

Me in my happy place in the woods on a family walk. [WL Gibson.]

The first recognition that the properties of a material are related to its microscopic structure—what engineers today call "structure-property relationships"—was made by the seventeenth-century scientist and artist Robert Hooke. His remarkable drawings for his book *Micrographia*, published in 1665, were the first to illustrate what insects, plants, and other materials look like under a microscope. When Hooke looked closely at cork for the first time, he saw an array of self-contained "microscopical pores" that, he realized, explained "the true and intelligible reason for all the *Phenomena* of Cork"—that is, its buoyancy, compressibility, and impermeability to water: the properties of cork are related to the structure of its microscopic pores (figure 0.7).[2]

Graduate research tends to be quite specialized. I focused on the mechanical properties of materials with a porous, cellular structure: engineering honeycombs, with their hexagonal cells, and foams, with their polyhedral cells; both are of interest for lightweight structural components, impact protection devices, and other applications (figure 0.8a). Since then, over four decades, my research has focused on how the microscopic structure of honeycombs and foams—the geometry of their cells, their porosity, and the solid they are made from—determines their strength and other mechanical properties. The work of my research group and our collaborators has included projects on natural cellular materials in plants (e.g., cork, wood, bamboo, iris leaves, and grass stems) and animals (e.g., spongy bone, lung alveoli) (figure 0.8b). I sometimes think that I have combined my father's interest in engineering with my mother's love of natural history.

When I first started teaching classes on the mechanical behavior of materials at MIT, I focused on the sorts of thing you would expect: deriving equations and making use of those equations in engineering applications. But in my last ten years of teaching (I retired in 2022), I included more examples from nature in my classes: how the cellular structure of wood gives rise to different strengths along and across the grain, how the internal structures of cattail and iris leaves make them lightweight without sacrificing strength, and many more. I also added stories about the people who developed the ideas that I taught, because it engaged and entertained students. For instance, when we studied Hooke's law, relating the stretch of a spring to the force applied to it (double the force, double the stretch), I also introduced them to Hooke's extraordinary drawings in *Micrographia*.

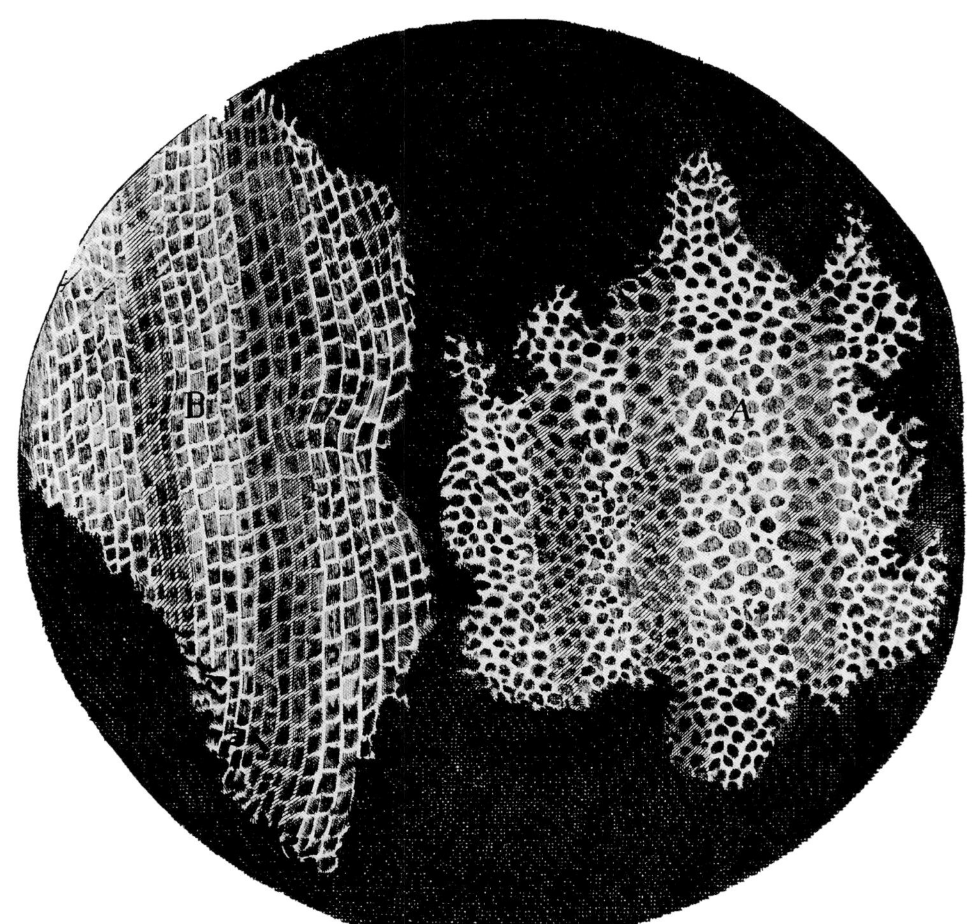

FIGURE 0.7

Robert Hooke's drawing of cork, showing two perpendicular cross sections, from his book *Micrographia*. [From Gibson and Ashby (1997), © Cambridge University Press. Reprinted with permission.]

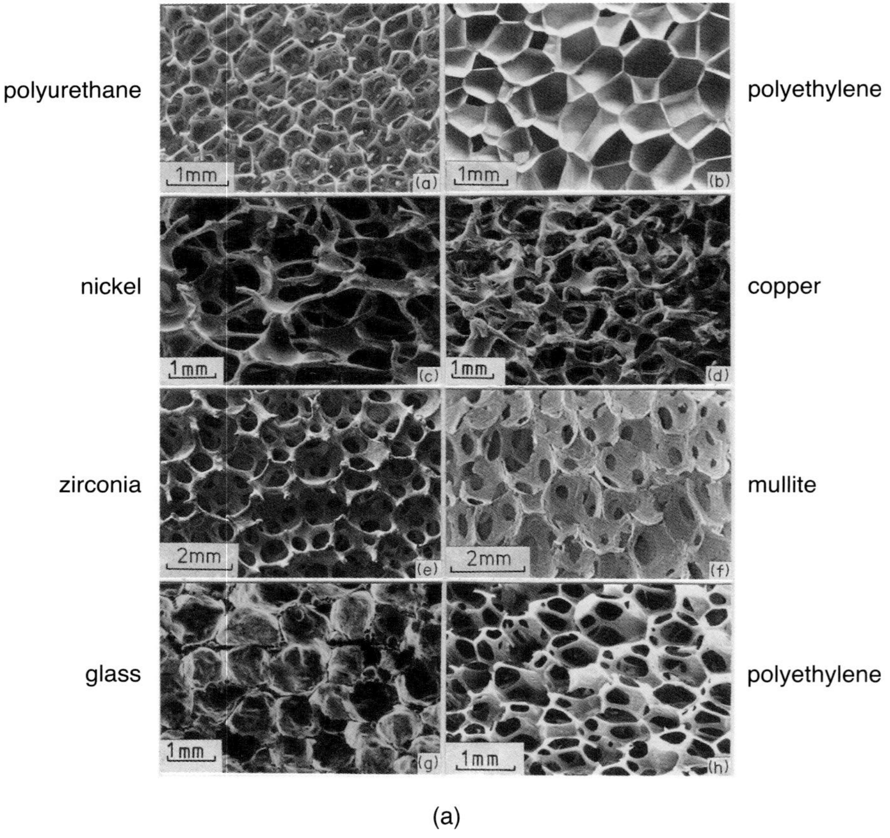

FIGURE 0.8

Scanning electron microscope images of (a) engineering foams and (b) cellular materials in nature. [From Gibson and Ashby (1997), © Cambridge University Press. Reprinted with permission.]

cork
balsa
sponge
spongy bone
coral
cuttlefish bone
leaf
stem
100 µm
(a)
100 µm
(b)
200 µm
(c)
1 mm
(d)
2mm
(e)
1mm
(f)
50 µm
(g)
250µm
(h)

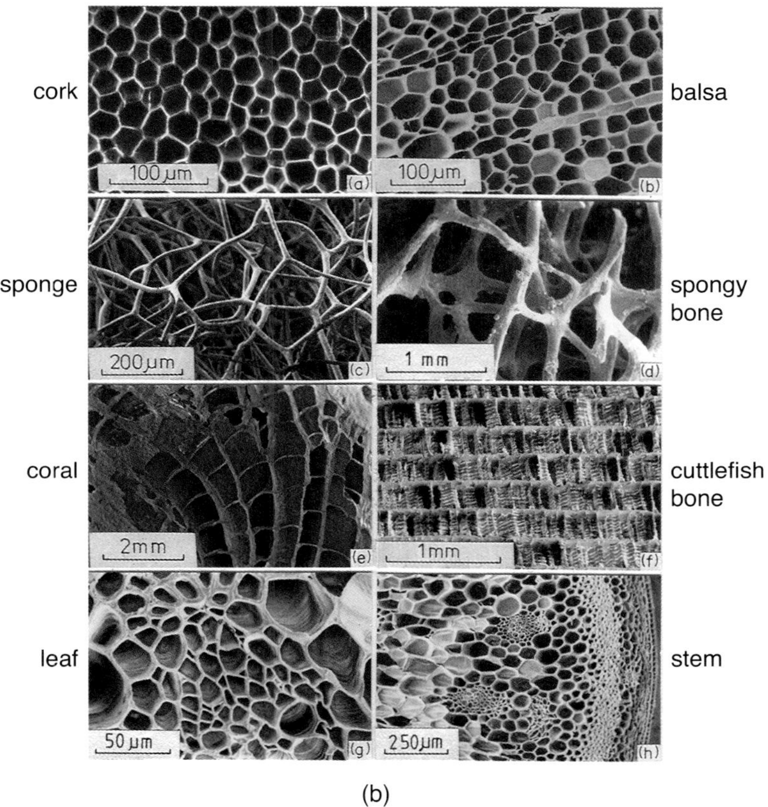

(b)

When I moved to Massachusetts in 1984 to take up my faculty position at MIT, I knew I was going to be spending a lot of time building up my research group and teaching. I also knew that I was going to need an occasional escape to nature to get out and relax. So I bought a small book, *The AMC Guide to Country Walks near Boston within Reach by Public Transportation*, by Alan Fisher, which soon became my bible for getting outdoors and exploring local conservation areas. Guided by Fisher, I discovered the nature sanctuaries of the Massachusetts Audubon Society, one of the oldest birding and conservation groups in the country. And for nearly 40 years, with a series of dogs (Sandy, a yellow labrador; Tobes, a chocolate [Toblerone] labrador; and Maddie, a cream-colored labradoodle), I've walked regularly in the green spaces in my neighborhood. These include the Forest Hills Cemetery, one of the early garden cemeteries, founded in the mid-nineteenth century to provide a parklike setting for burials; the Arnold Arboretum, a 281-acre collection of temperate woody plants that was the first public arboretum in North America; and the mile-and-a-half footpath around Jamaica Pond, a glacial kettle pond. Through it all, I've grown more interested in birds. On my walks now, I keep an eye out for them and write down what I see in a small notebook.

There's lots to see.

At the Forest Hills Cemetery early one mid-February morning, I was startled to see a pair of red-tailed hawks flying low, one carrying a substantial stick, perhaps 20 inches (0.5 m) long and an inch (25 mm) in diameter, in its beak. They headed toward a cluster of tall pines that stood near the grave of the playwright Eugene O'Neill. About a week later, I saw the nest, high up in one of the pines, supported between the trunk and several branches. After that, I checked in regularly on the birds' progress. A couple of weeks later, I spotted one of the hawks with another stick in its beak, flying onto the nest; in early March, I saw two hawks, possibly my pair, up in a tree by the cemetery pond, squawking and mating; and in early June I saw a nestling, near fledging, hopping about in the nest. The following winter, I noticed that the nest had been taken over by a pair of great horned owls (figure 0.9), so I continued to monitor it. At the end of April, I saw three owlets out of the nest, two of them hop-flying on a branch while the third—with developing flight feathers on its wings but fluff still on its breast—perched immobile on a lower branch of the tree. I did a quick scan of the vicinity and found one of the adults observing them all from a neighboring tree. Hawks and owls, young and old—what a thrill each time to see them!

FIGURE 0.9

A great horned owl nest, with one of the adults and an owlet, at the Forest Hills Cemetery in Jamaica Plain, MA. [Photo: Robert Mayer.]

Another year, a pair of great blue herons nested in one of the pines on the tiny island in the pond at the cemetery, near the grave of the poet e.e. cummings (whose stone is inscribed EDWARD ESTLIN CUMMINGS). My wife, Jeannie, and I loved going to see them in the evenings after dinner. Often we'd hear the young even before we could see them; waiting to be fed, they would hold their bills up vertically, each one clapping its upper and lower bills together, making quite a racket. One evening, at dusk, I saw a Boston bicycle-police officer sitting on a large rock near the pond, patrol bike lying on the ground beside him, watching the nestlings. As I joined him, he told me he was concerned that the adults had not yet returned for the night. Boston's finest sometimes even watch over baby birds.

I love it when migrating ducks appear at Jamaica Pond in the fall. It's quite a parade. The ever-so-elegant hooded mergansers usually arrive first, early in November, and stay until the pond freezes over. They're soon joined by a stream of others: ring-necked ducks and ruddy ducks, the latter almost comical with their stiff upward-pointing tails and abrupt dives down into the water; a few common mergansers; the occasional red-breasted merganser; and, from time to time, wood ducks, the males with delicate feathers seemingly painted on with a fine brush. Joining the fray are pied-billed grebes, small and energetic, diving over and over again in the water, and northern shovelers, skimming the surface of the water for aquatic invertebrates with their shovel-like bills. And once in a rare while, a lone loon. It's all delightful.

Most exciting for me are the bald eagles that sometimes appear in the winter. One year I watched as one took off from a tree at the shoreline and went after a gull over the pond. The gull darted about, this way and that, evading the eagle, but the eagle matched every swerve and dive, following close behind it. Three crows soon arrived and began mobbing the eagle as it chased after the gull. Back and forth, up and down the pond they all went—until eventually the eagle gave up and landed on a pine tree across the pond. It was a magnificent display. When I told some friends about it a day or two later, they said that their dog had recently found a gull wing in the leaf litter by the pond footpath. Perhaps left over from a successful eagle attack.

The true *National Geographic* moment, though, came when Jeannie and I were on our way to get groceries one bitterly cold February day, when the pond was nearly completely frozen over. Driving beside the pond, I spotted a bald eagle perched high up in a sycamore tree at the northwest corner of the pond—with a hawk harassing it! I quickly pulled over and jumped out of the car to get a better look. Ignoring the hawk, which gave up and flew off after a few minutes, the eagle had its eyes on a small group of coots in a tiny circle of open water just in front of us, about 50 feet

(15 m) away. Suddenly, the eagle swooped down, talons out, going after the coots. Just as it got close, the coots dove, then popped back up as soon as it had passed by. The eagle tried several more times before giving up and flying off to the other end of the pond. I could have watched the show all day.

One of my most exciting and memorable moments of birding came a bit farther afield, at the Parker River National Wildlife Refuge, on Plum Island, about an hour's drive north of Boston, where I saw Norm Smith, of the Massachusetts Audubon Society, release a snowy owl into the wild. The owl had been captured, with appropriate permits, at Logan Airport, which sits on a strip of land just offshore from downtown Boston (figure 0.10). Surprisingly, the winter landscape around Logan, with its dried grasses and scrub, looks enough like the summer Arctic tundra to attract the largest known concentration of snowy owls in the northeastern United States. They usually arrive in early November and depart by early April; while at the airport they prey on small mammals as well as other birds. But because the owls are a hazard to aircraft, Norm captures as many as he can and releases them on the beaches north and south of the city. Most years, Norm captures eight or nine, but during the winter of 2013–2014, when there was an irruption of snowy owls in

FIGURE 0.10

A snowy owl at Logan Airport in Boston, MA. [Massachusetts Audubon Society, photo: Norm Smith. Reprinted with permission.]

the eastern United States, he trapped and relocated 121. Ever since learning about this, I've been on the lookout for a snowy owl whenever I fly into or out of Logan in the winter.

When I tell my students that I go birding in Boston, they're often surprised, because they feel there aren't any interesting birds to see in the city. I tell them that if you think you won't see anything, and you don't look, then you won't see anything. Me, I'm always on the lookout. And there are *always* lots of birds to see.

When I was a graduate student working on the mechanics of foams, a colleague mentioned to me that he had heard that woodpeckers have a foam-like material between their skulls and brains to protect their brains from the impact of pecking. And protection certainly would seem to be something they need: they peck over and over, up to 20 times a second, and hard enough to make quite a racket. At the time, I was focused on my doctoral thesis, but I kept this in the back of my mind, thinking that one day I would look into it. After all, I love birds and specialize in the mechanical properties of foams, so how could I resist?

Thirty years later, I came back to it, thinking it would be a great way to combine my interests in engineering and birds. I started by reading everything I could on the subject, and this led me to a 1976 study by a group of California neurologists who had dissected a woodpecker's head to better understand how woodpeckers avoided brain injury. They, too, had heard the special-foam theory, but their dissection disproved the theory. They found no foam at all.

Now *that* was intriguing to me. If there's no foam inside the skull of a woodpecker, then how do woodpeckers avoid brain injury? I was hooked.

After studying the question, eventually I recognized that because woodpeckers have such tiny brains, they don't need the kind of protection that larger animals would need. Force is equal to the mass of an object times its acceleration (or deceleration, if slowing down). For the same deceleration on impact, a smaller mass produces a smaller impact force. An acorn woodpecker brain has a mass of 2 grams (a little less than a US penny) while the human brain has a mass of 1,400 grams (about 3 pounds); for the same deceleration on impact, the force on the human brain is 700 times that on the woodpecker brain. (Yikes!)

I wrote a scientific paper about this, going into more detail, and started giving talks about it—first to academic audiences, and then, in a slightly different version, to birding groups.[3] The talks to birders were more fun, and I began to imagine writing a whole book for birders about how birds work using my background in materials science and mechanics, drawing on my love of birds, engineering, and teaching. It is my hope that this book will give birders and others interested in natural history a new perspective on the wonder of birds.

HOW TO READ THIS BOOK

Engineers use sketches, plans, and images to help explain their ideas. I would have found it impossible to teach my engineering classes at MIT without such visualizations, and so, as you'll discover, I've adopted that same approach in writing this book, which relies heavily on illustrations. Please spend time with them, as they are not only helpful in making clear the ideas that I'm explaining but are also simply beautiful in their own right.

I should also note that I've written this book for readers who have no special background in science or engineering. The only prerequisite for entry is an interest in birds and natural history. Engineering analysis can be highly mathematical, of course, but the rule I've made for myself in writing this book is to put any calculations into boxes and to restrict the mathematics to addition, subtraction, multiplication, division, and exponents. If those calculations interest you, great, but you can read the book without going into the details of the box calculations.

A final note: although the book is organized into chapters, each section within the chapters can be read on its own. So if you're interested in what makes feathers so water-repellent, for example, you can read that one section without needing to read the entire chapter.

REFERENCES

Fisher A (1976). *The AMC Guide to Country Walks near Boston within Reach by Public Transportation*. Appalachian Mountain Club.

Gibson LJ (2006). Woodpecker pecking: how woodpeckers avoid brain injury. *Journal of Zoology* **270**, 462–465.

Gibson LJ and Ashby MF (1997). *Cellular Solids: Structure and Properties*. Second edition. Cambridge University Press.

Hooke R (1665). *Micrographia*. Royal Society of London.

May PRA, Fuster JM, Newman PA, and Hirschman A (1976). Woodpeckers and head injury. *Lancet* **307**, 454–455.

1

FANTASTIC FEATHERS: COLOR AND SOUND

Everyone likes birds. What wild creature is more accessible to our eyes and ears, as close to us and everyone in the world, as universal as a bird?

David Attenborough

A female broad-tailed hummingbird is perched on a tree branch in a high mountain meadow in the Colorado Rockies. A male, flying high above her, plunges in a death-defying, U-shaped courtship dive over the female's head, dropping a hundred feet and reaching 50 miles per hour. At the bottom of his dive, as he passes the female and finally pulls upward, his iridescent throat feathers flash from magenta to black and simultaneously his tail feathers snap together to produce a loud buzzing sound. The female is dazzled. At least that's what he hopes.

Mary Caswell (Cassie) Stoddard, of Princeton University, and her postdoc, Ben Hogan, have conducted video and audio analyses of this remarkable courtship performance. "You really have to see it to believe it," she says. "The first time I saw it, I almost fell over." Stoddard and Hogan have shown that at the critical moment, at the base of the dive, in front of the female, the male synchronizes the flashing of his throat feathers with the buzzing of his tail feathers to within 300 milliseconds—as fast as the blink of an eye.

How do they do it? What makes some feathers iridescent while others are not? How do feathers produce sound? And how do other feathers *suppress* sound, allowing owls to fly almost silently as they approach their prey? For that matter, how do the feathers on owls' facial ruffs focus sound toward their ears, enabling them to hunt by hearing alone?

The answers to all these questions lie in the microscopic structure of contour feathers—the outer feathers that cover a bird's body.

CONTOUR FEATHER STRUCTURE

Like all feathers, contour feathers are modified hairs, and like hair they grow from follicles in the skin. Like mammals' hair, horn, and hooves (and reptiles' scales), they are made of keratin. Unlike hair, however, they have stunningly complex structures.

The vane of a contour feather is a branched structure, with *barbs* branching in parallel off the shaft of the feather, and *barbules* branching in parallel off the barb shafts (figure 1.1). A single barb is made up of its shaft as well as the attached barbules.

The barbules on the side of the barb shaft closest to the base of the feather have grooved cross sections in the shape of a comma, while those on the other side have tiny *hooklets* that latch into the grooves of the barbule on the adjacent barb, producing an interlocking structure. When a bird preens, running its bill along the feather vane from base to tip, it engages the hooklets with the grooves, smoothing the vane.

Barbules are too small to be seen with the naked eye, and to measure them we have to use the unit of the micron (μm), which equals one thousandth of a millimeter. The barb shafts of water birds' contour feathers are typically 20–70 microns wide. Barbules from flight feathers are even smaller, at about 10 microns wide, and hooklets are smaller still, measuring just a few microns across. (Human hair, by comparison, is typically about 50 microns in diameter.)

We can see even more detail of the structure of a contour feather when we look at scanning electron micrographs (photographs taken by a microscope) (figure 1.2). Looking down on top of the feather vane (figure 1.2b), we can see the feather shaft, with the barbs branching off it and the barbules branching off the barb shafts. Looking at cross sections of the feather (figure 1.2c, d), we can see that both the shaft of the feather and the shafts of the barbs have a porous, foamy core. We can also see the interlocking of the hooklets and grooves of adjacent barbules (figure 1.2c, d) and—in one of my favorite images—a single barb and its attached barbules, with their grooves and hooklets (figure 1.2e).

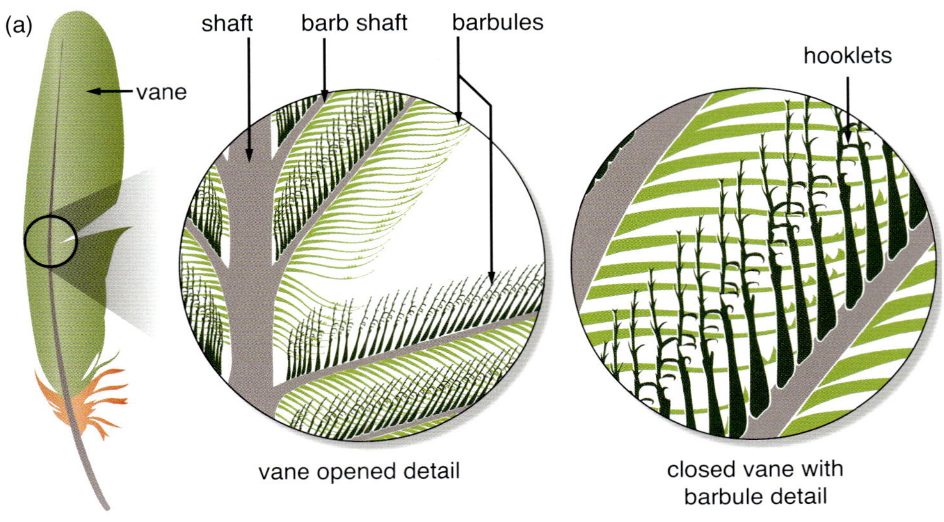

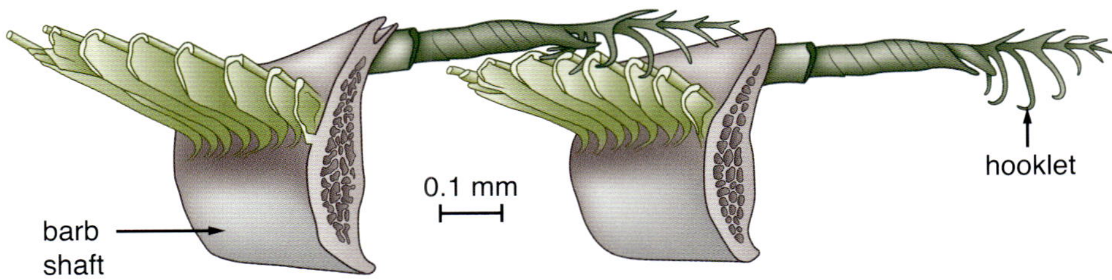

FIGURE 1.1

The structure of a contour feather, one of the outer feathers on a bird, showing its branched structure. [(a) Illustration by Andrew Leach, © Cornell Lab of Ornithology; (b) adapted from Lucas and Stettenheim (1972).]

FIGURE 1.2

(a) A mallard contour feather. (b, c, d, e) Scanning electron micrographs of a mallard contour feather: (b) looking down on the vane, (c, d) cross sections, and (e) a single barb, with its attached barbules. The scale bars are in microns (µm: 1 µm = 0.001 mm); a human hair is roughly 50 microns across. [Scanning electron microscope images taken by Dr. Isaac Cabrera.]

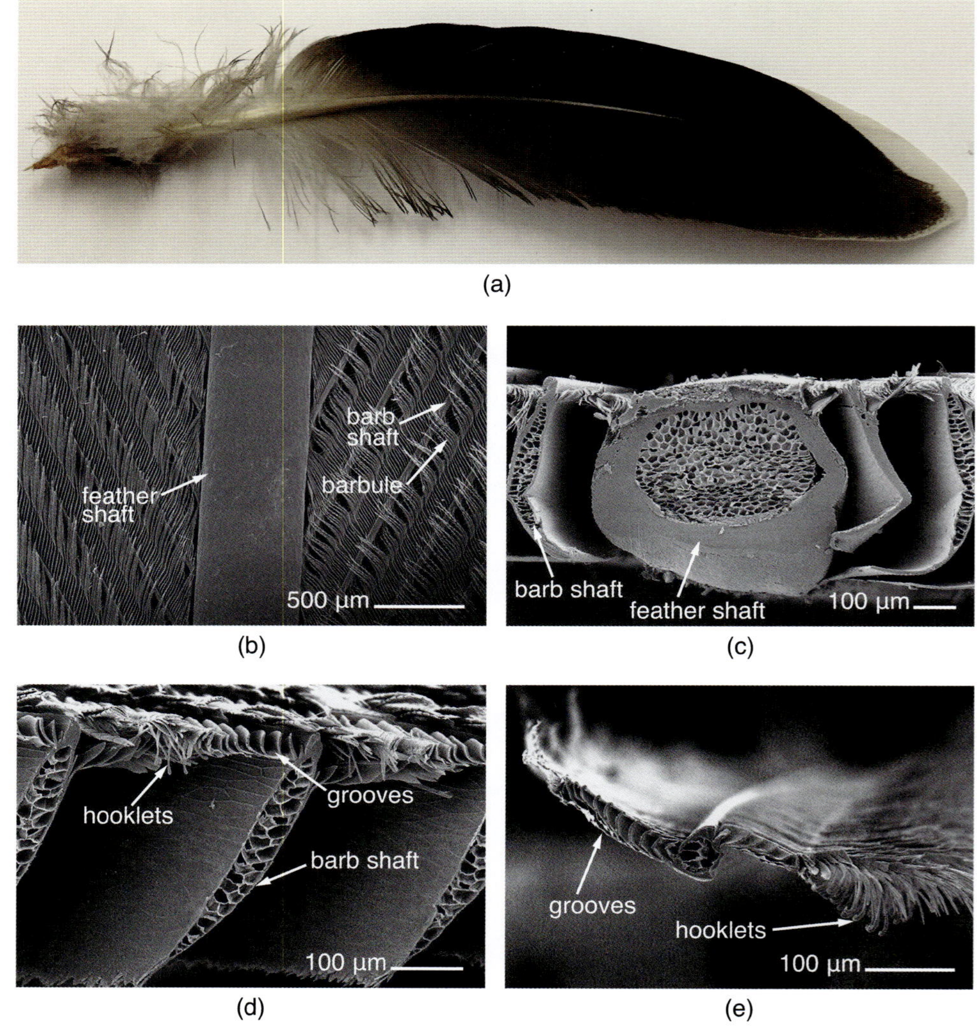

diameter of human hair ~50 microns = 50 µm

In this chapter and the next, we'll see how the structure of feathers, particularly at the microscopic level, gives rise to their remarkable versatility: their broad palette of colors, including iridescence, their ability to create, suppress, and collect sound, their water repellency, and their ability to insulate from the cold.

FEATHER COLOR

Birds are among the most dazzlingly colorful creatures on the planet. That's what often first draws people into birding. The American flamingo is pink all over, except for black edges on its wings. The male wood duck exhibits a combination of colors that seems as though it just shouldn't be possible: a green head crest, a black face, a white bridle on the front of the neck, a speckled-chestnut breast, a slatey-gray back, and a buff belly. Hummingbirds shimmer mesmerizingly with iridescent greens, violets, and rusts. You can find warblers with all manner of delightful patterns: yellows with grays or blues; black and white streaks; chestnut sides; striped heads.

But what is color?

The color we see comes from light that is reflected from a surface. Light can be thought of as a wave, like a wave on the ocean. The shape of a wave is characterized by the distance between the crests, the *wavelength*, and the wave's height, or *amplitude* (figure 1.3). Different colors correspond to light of different wavelengths, and the spectrum of light visible to the human eye ranges from violet (with a wavelength of about 0.4 microns) through blue, green, yellow, and orange to red (with a wavelength of roughly 0.7 microns). Beyond the spectrum of light visible to the human eye are ultraviolet light (with wavelengths shorter than that of violet) and infrared light (with wavelengths longer than that of red).

In daylight, all the visible colors of light are present in roughly equal intensities, and you can see the full spectrum when you shine light through a prism. When light hits a surface, some of the wavelengths are absorbed and some are reflected—and the wavelengths of the reflected light produce the colors that we see.

We perceive color through three types of cones in the retina at the back interior surface of the eye: "blue" cones, which are sensitive to short wavelengths of light; "green" cones, which are sensitive to medium wavelengths; and "red" cones, which are sensitive to long wavelengths (figure 1.4). In the human eye, these three types of cones cover the entire spectrum of colors that we see.

FIGURE 1.3

Color: light of different wavelengths. (a) The wavelength is the distance between crests of the wave. (b) Light of different wavelengths gives different colors. (c) The wavelengths of the visible spectrum, from violet to red. 0.5 µm is roughly 1/100 the diameter of a human hair. [(b, c) After https://science.nasa.gov/ems/09_visiblelight]

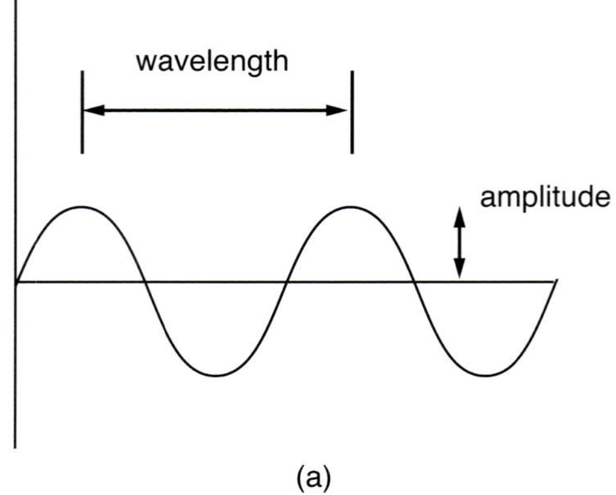

(a)

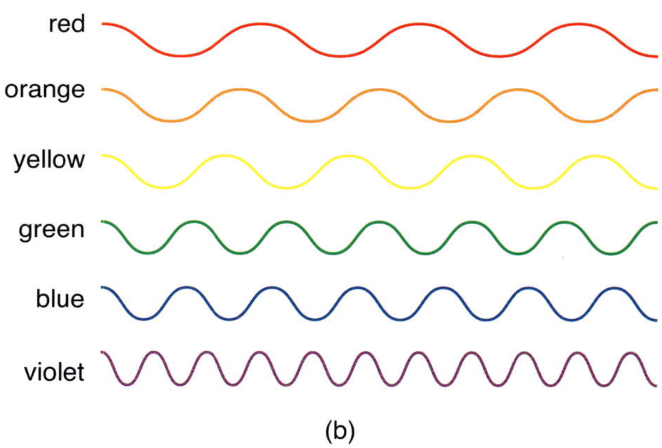

(b)

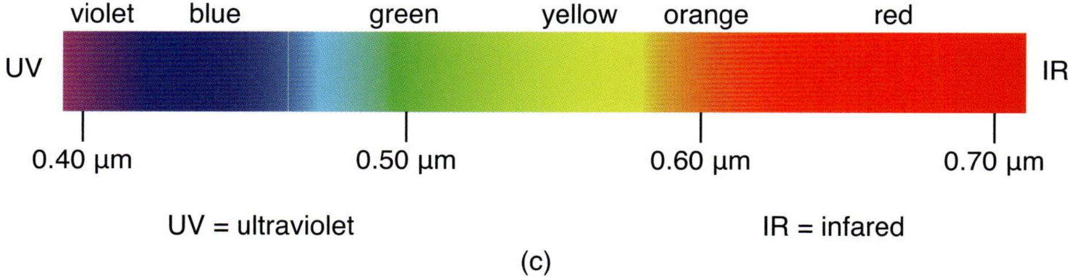

(c)

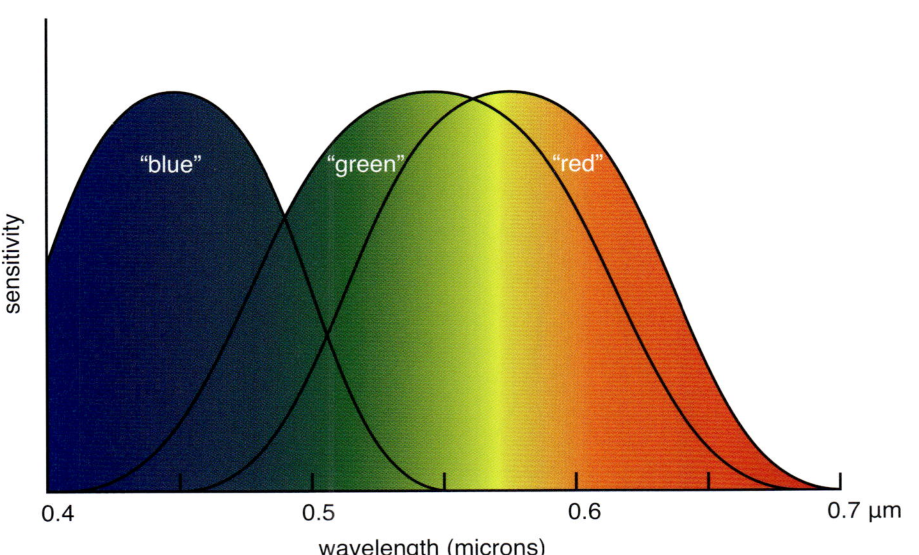

FIGURE 1.4

The wavelengths of light (and corresponding colors) that the three cones in our eyes are sensitive to. [After http://hyperphysics.phy-astr.gsu.edu/hbase/vision/colcon.html]

Blue cones are most sensitive to blue light but also detect colors with wavelengths near that of blue. When blue light hits the eye, the blue-sensitive cones send a signal to the brain that is interpreted as the color blue. The same principle holds for green and red cones. Colors in between blue, green, and red stimulate more than one type of cone, and what we perceive depends on the level of stimulation in each type of cone. For instance, the yellow feathers of a yellow warbler strongly stimulate both the green and red cones, making us see yellow.

Diurnal birds, active during the day, are more sensitive to color than people, because they have a fourth type of cone in their retinas that is sensitive to ultraviolet light (figure 1.5). This means that they can see colors that we can't even conceive of. What color do you get when you mix ultraviolet and green, for example, or ultraviolet and red? We'll never know, but these birds do.

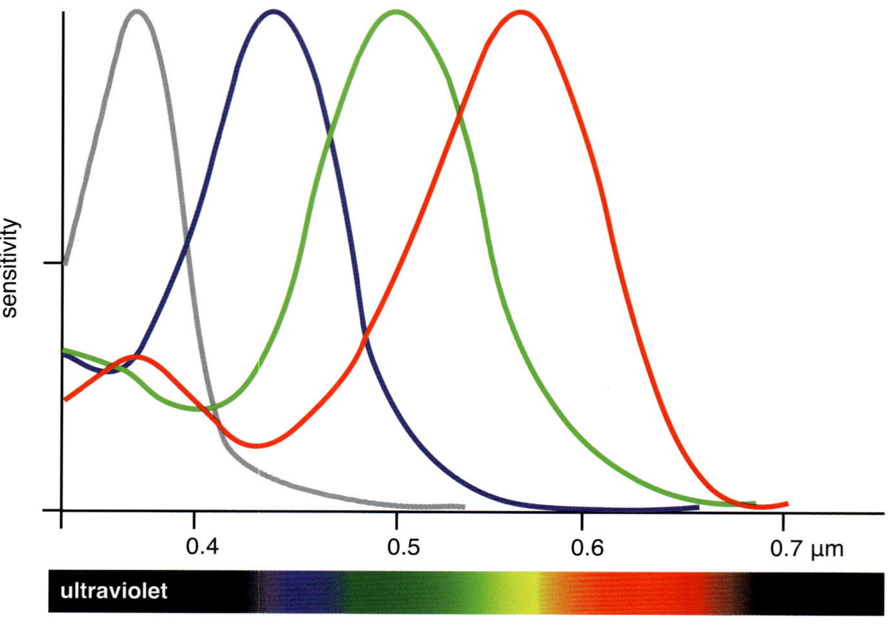

FIGURE 1.5

The wavelengths of light (and corresponding colors) that the four cones in estrildid finches' eyes are sensitive to. Other diurnal birds have similar sensitivities. [After Hart et al (2000) and https://en.wikipedia.org/wikiTetrachromacy#/media/File:BirdVisualPigmentAbsorbance.svg Shyamal L, public domain.]

And the distinctions matter to them. Kevin Burns, at San Diego State University, and Allison Shultz, at the Natural History Museum of Los Angeles County, have shown that some bird feathers have these combined ultraviolet colors that are visible to birds. For instance, they have shown—by measuring the wavelengths of light reflected from feathers and applying a model of bird vision that accounts for the sensitivities of their four cones—that in nearly all species of tanagers, there are differences in the colors of the feathers between males and females that the birds can see, even though the feathers look the same to us.

What about nocturnal species of birds? Among the species of owls that have been studied, none have been found to have ultraviolet-sensitive cones. That's not surprising: in primarily nocturnal species there's no advantage to having a more refined sensitivity to light.

The cones in birds' eyes have another remarkable feature that heightens birds' sensitivity to color: colored oil droplets that absorb some of the incident light. The droplets limit the bands of wavelengths of light (or colors) that reach and stimulate the cones. Within these narrower bands, the cones can then distinguish smaller variations in color than human eyes can, allowing them to detect, for instance, green insects on slightly lighter or darker green leaves.

Studies of feather color typically rely on collections of birds in natural history museums. You might think that natural history collections are dusty, musty, and outdated, but in fact they're invaluable for research studies. Throughout this book, we'll see how the research collections in natural history museums have been used to study everything from feather color to egg shape to the ability to fly. They've even been used to study genetic variability in populations over time, because they often include specimens collected decades or even centuries ago. Shultz refers to such specimens as "time machines."

PIGMENTS

Most feathers are colored by pigments that absorb some wavelengths of light while reflecting others, giving the color we see. The most common pigment in feathers is melanin, present in at least some feathers of nearly every bird (melanin is also present in human hair and skin). There are two types of melanin: eumelanin, consisting of black granules, and pheomelanin, consisting of light-brown or reddish-brown granules. In dense concentrations, eumelanin produces the blacks of blackbirds, cormorants, and ravens. At lower concentrations, it produces gray—the lower the concentration, the lighter the shade of gray. The different shades of gray or black

FIGURE 1.6

Pigments in feathers. [All Alamy.]

eumelanin
black, gray

(a) herring gull

pheomelanin
reddish or light brown

(b) cinnamon teal

carotenoids
red, yellow

(c) scarlet tanager

carotenoids
red, yellow

(d) red-eyed vireo

turacin and turacoverdin
red, green in turacos

(e) Knysna turaco

psittacofulvin
yellow, orange, red in parrots

(f) red-and-green macaw

wings of many gulls (figure 1.6a) reflect variations in the concentration of eumelanin pigment in the feathers. Similarly, dense concentrations of pheomelanin produces the reddish-brown of the cinnamon teal (figure 1.6b), and variations in the concentration of pheomelanin and in the proportion of pheomelanin to eumelanin produce many shades of brown to rusty-red.

Iron-based porphyrin pigments are found in rusty-red and brown feathers in roughly half the orders of birds, particularly in owls and bustards. Iron-based porphyrins are unstable in ultraviolet light, causing them to decay when exposed to the sun; they are only found in relatively new feathers. They always appear with pheomelanins, which create the long-term color of the feathers. Porphyrins also produce reddish and brown markings on bird eggs.

Carotenoid pigments, the second most common pigments in feathers after melanins, produce red, orange, or yellow. Carotenoid pigments are made only in plants, algae, and fungi; the only ways for birds to obtain them are either by eating plants, algae, or fungi containing them, or by eating something else that does. Flamingoes, for instance, get their striking pink color from carotenoids in the algae and brine shrimp that they eat; if their diet lacks carotenoids, their feathers are pale. As with the melanins, variations in the concentration and proportion of red and yellow carotenoid pigments produce colors ranging from the intense red of the scarlet tanager to the pale yellow of the red-eyed vireo (figure 1.6c, d).

There are several other, less common pigments that occur in the feathers of a limited number of species of birds. The vivid reds and greens of turacos in Africa (figure 1.6e) are produced by two copper-based porphyrin pigments, turacin and turacoverdin, respectively; these two pigments are only found in turacos. Psittacofulvins, unique to parrots, give them their bright yellow, orange, and red colors (figure 1.6f).

STRUCTURAL COLOR

The pigment palette in feathers includes black, gray, red, yellow, orange, brown, and green, but not blue; there are no blue pigments in feathers. So where do the blues of blue jays, eastern bluebirds, belted kingfishers, and cerulean warblers come from? Blue, violet, and ultraviolet are structural colors, resulting from the interaction of light with microstructural features within the feather barb that are similar in size to the wavelength of light.

How does this work?

We already saw that in a contour feather, the barb shaft has a thin outer shell of solid keratin (that is transparent) and a foam-like core with pores that measure tens of microns across (figure 1.2d). In the barb shafts of blue feathers, like those of the eastern bluebird (figure 1.7), there is a spongy keratin layer between that outer shell and foamy core—a layer with even tinier pores that are only a few tenths of a micron in size. These pores are underlain by melanin granules.

When Rick Prum was at the University of Kansas (he's now at Yale University), he and his colleagues found that the submicron pores in the spongy keratin layer of blue feathers can be either channel-like, as, for instance, in the eastern bluebird, or roughly spherical. The size of the pores is such that wavelengths of light corresponding to blue reflect from the spongy keratin layer and are visible, while longer wavelengths (greens, reds) pass through the pores and are absorbed by the melanin granules beneath the spongy keratin (figure 1.8). By measuring the pore sizes in the spongy keratin layers of blue feathers, Prum and his colleagues have been able to predict the exact hue of a blue feather (measured from the wavelengths of the reflected blue light). The slightly different blues of blue jays, eastern bluebirds, belted kingfishers, and cerulean warblers, it turns out, are all produced by slightly different pore sizes in the spongy keratin layers of their blue feathers.

Geoff Hill, an ornithologist at Auburn University, tells the story of asking a graduate student in an oral exam how she would remove the color from two different feathers: a red feather from a scarlet ibis, and a blue feather from a blue-bellied roller. The student responded that the red feather is pigmented with carotenoids, so she would remove them chemically. But she came up with a very different plan for the blue feather: knowing that destroying the microscopic structure of the feather would destroy its colors, she said she would smash it with a hammer!

Ultraviolet color in feathers is usually a structural color, produced by the same mechanism as blue, but with slightly smaller pores in the spongy layer (since ultraviolet has a smaller wavelength than blue light). No pigment has yet been found that reflects solely UV light, but some carotenoid pigments reflect ultraviolet along with other wavelengths of light.

Blue structural color can be combined with yellow or red pigment to produce greens and purples. For instance, green feathers in songbirds, such as the black-throated green warbler, are produced by a combination of blue structural color with yellow carotenoid pigment; the vivid greens of parrots arise from a combination of blue structural color and yellow psittacofulvin pigment, and the purple feathers on the head of the blossom-headed parakeet combine structural blue from the barbs with red pigment in the barbules.

(a) eastern bluebird

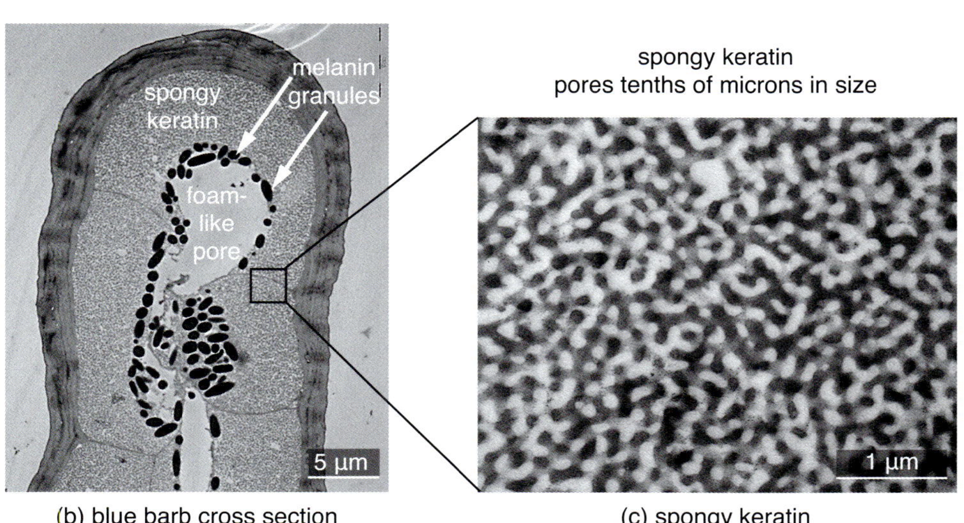

(b) blue barb cross section

(c) spongy keratin

FIGURE 1.7

(a) Eastern bluebird. (b) Cross section of a barb shaft from a blue feather from the back of an eastern bluebird. The foam-like pores seen in all bird barb shafts are roughly 10 µm across. The spongy keratin seen in blue barb shafts has pores tenths of a micron in size. (c) Higher-magnification electron microscope image of the spongy keratin. [(a) Alamy; (b, c) courtesy of Rick Prum.]

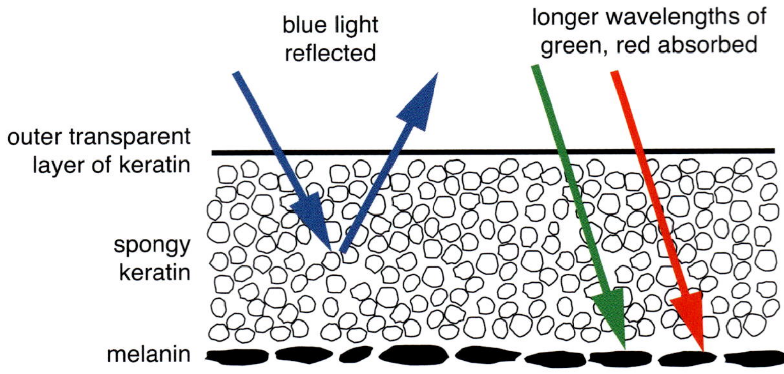

FIGURE 1.8

Schematic of a barb shaft in a blue feather. Blue wavelengths are reflected by the spongy keratin layer, while longer wavelengths (green, red) are absorbed by the melanin layer below. [After Hill (2010).]

Iridescence, too, is a structural color. Striking displays of iridescence appear in the feathers of many hummingbirds, the scales of some butterfly wings, the outer cuticle of beetles, and a thin layer of oil on a puddle of water (figure 1.9). One of the defining characteristics of iridescence is that the apparent color changes with the angle of viewing: for instance, the throat of the ruby-throated hummingbird turns to a deep olive green when viewed from the side instead of from the front (figure 1.10).

People have been fascinated by iridescence for centuries. The word itself comes from the Greek goddess Iris, who was thought to be the personification of the rainbow; Iris was also the messenger between the gods and the natural world.

In 1665, Robert Hooke wrote memorably about iridescence in peacock and duck feathers in his book *Micrographia*, noting of certain feathers that "by means of various positions, in respect of the light, they reflect back now one colour, and then another, and those most vividly."

Hooke correctly suspected that iridescence arose from some microscopic structure within the feathers, but his microscope wasn't powerful enough to detect the tiny features that produce iridescence. In fact, those features are so small, less than the wavelength of visible light, that they could not be detected until after the invention of electron microscopes.

(a)

(b)

(c)

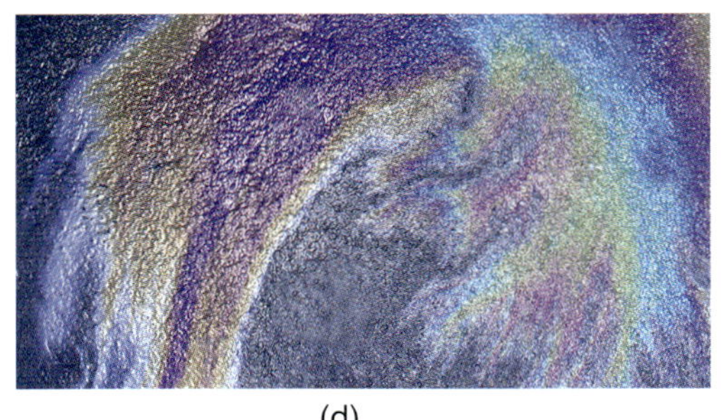

(d)

FIGURE 1.9

Iridescence in (a) the throat feathers of a calliope hummingbird, (b) the wings of a blue morpho butterfly, (c) the cuticle of a beetle, and (d) oil on a puddle of water. [All Alamy.]

FIGURE 1.10

The throat feathers of a ruby-throated hummingbird appear in different colors when viewed from different angles, the defining characteristic of iridescence. [Both Alamy.]

The simplest way to understand iridescence is to consider a thin film of oil on water (figure 1.11). When light hits a thin layer of oil on water, some of the light is reflected off the surface of the oil while the rest is transmitted through the oil. A light ray travels through the oil at a different angle than through the air: it refracts. You can see an example of this if you hold a straight stick in water: the stick appears to bend at the interface between the air and the water due to refraction, caused by the different speed of light in air and water.

Of the light that is transmitted through the oil, some reflects off the interface between the oil and the water and then travels back into the air above the oil. (Some, not shown in the figure, is also transmitted into the water.)

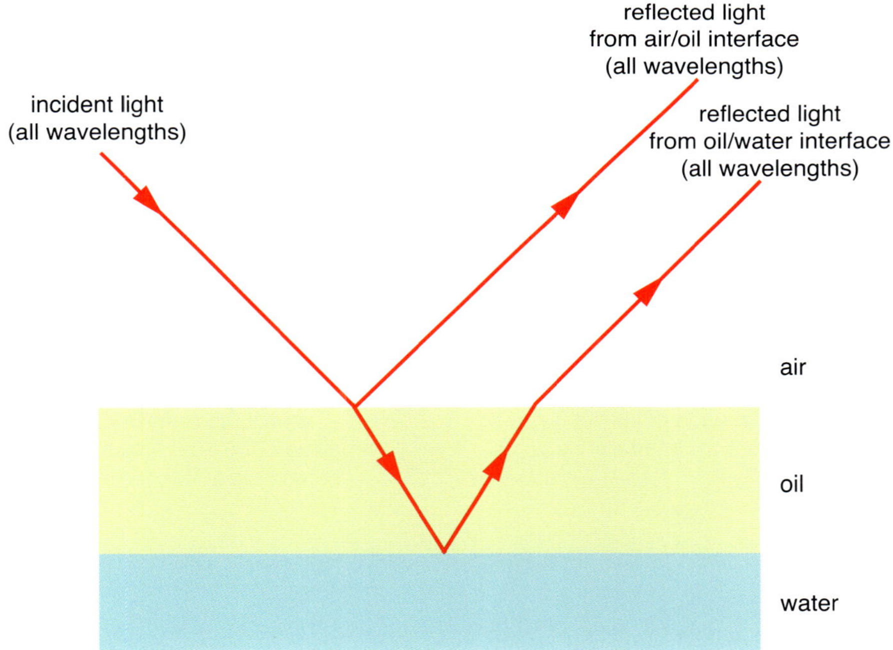

FIGURE 1.11

Light interacting with a thin layer of oil on water. [After https://en.wikipedia.org/wiki/Thin-film_interference#/media/File:Thin_film_interference.svg by Nicoguaro CC BY 4.0.]

Iridescence results from the interaction of the rays reflecting off the air-oil and oil-water interfaces. Two examples of how waves interact with each other are shown in figure 1.12. On the left, wave A and wave B have the same wavelength and amplitude, but the troughs of A align with the peaks of B (the waves are said to be "out of phase"). When added together, the two waves cancel each other out. On the right, wave A and wave B again have the same wavelength and amplitude, but this time the peaks and troughs of the two waves are aligned (the waves are said to be "in phase"). When added together they produce a wave with the same wavelength as A or B but double the amplitude.

In the oil-and-water example, the light ray passing through the oil layer and reflected off the lower oil-water interface travels a different distance than the light

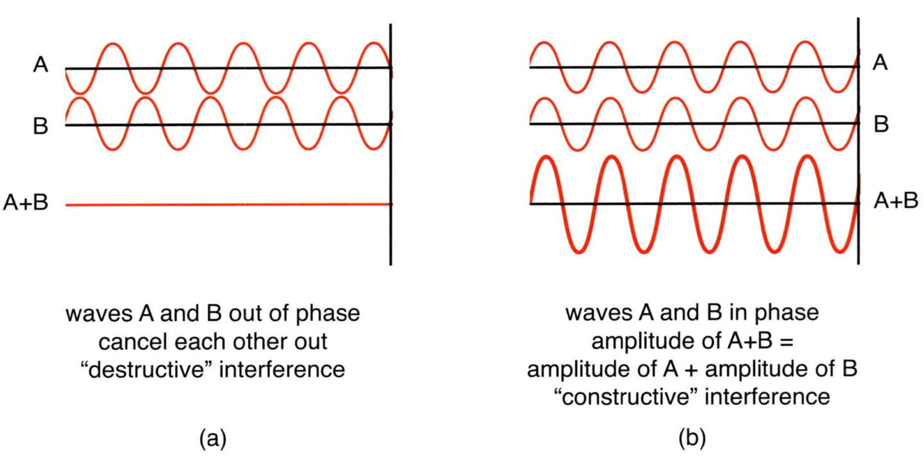

FIGURE 1.12

Two waves interacting. (a) The peaks of wave A coincide with the troughs of wave B; when added, the two waves cancel each other out (destructive interference). (b) The peaks of wave A coincide with the peaks of wave B; when added, the two waves reinforce each other to produce a wave with a larger amplitude (constructive interference). [After https://en.wikipedia.org/wiki/Wave_interference#/media/File:Interference_of_two_waves.svg by Haade vectorization: Wjh31, Quibik; CC BY-SA 3.0.]

ray reflected off the upper surface of the oil (figure 1.11). Each light ray is made up of all wavelengths of light. Most of the wavelengths of light from the two reflected waves (one from the upper air-oil interface and the other from the lower oil-water interface) will cancel each other out. But some interacting wavelengths will be in phase, with their peaks coinciding, increasing their amplitude and the intensity of their color. The color that you see corresponds to the wavelengths that are in phase.

Iridescent color changes with viewing angle because the distance that a wave travels through the oil depends on the angle of viewing (figure 1.13). At different viewing angles, the different distances that a light ray travels through the oil layer give different wavelengths of reflected light that are in phase and add up to give different intense colors.

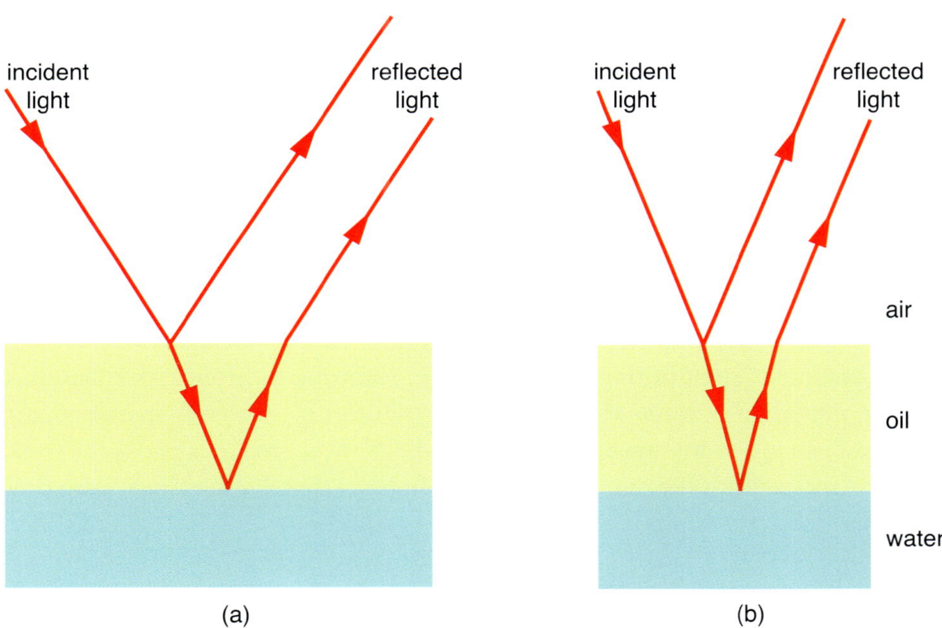

FIGURE 1.13

For different viewing angles, a light ray travels different distances through the oil layer. As a result, the wavelength that matches up (and constructively interferes) in the two light rays leaving the oil/air surface is different, and the viewer sees a different color. [After https://en.wikipedia.org/wiki/Thin-film_interference#/media/File:Thin_film_interference.svg by Nicoguaro CC BY 4.0.]

The iridescent bluish-violet feathers of the adult male satin bowerbird, found in eastern Australia, almost glow with brilliance (figure 1.14a). Stéphanie Doucet, Matt Shawkey, and Geoff Hill at Auburn University, in Alabama, and Robert Montgomerie, at Queen's University in Kingston, Ontario, found that the iridescence of its rump feathers arises from the microscopic structure of the barbules. An electron micrograph of the cross section of a barbule shows an outer shell of solid, transparent keratin, of roughly constant thickness, underlain by melanin granules; the top layer of melanin runs nearly parallel to the outer surface of the barbule (figure 1.14c, d). The average thickness of the outer, transparent keratin layer is 0.163 microns, about one-third of the wavelength of blue light.

The transparent outer shell of keratin is analogous to the thin oil layer in the oil-water example, while the melanin granules act as the water layer. Some of the light is reflected off the surface of the outer keratin shell, and some off the interface between the outer keratin shell and the top layer of melanin granules; most wavelengths of the two reflected light waves cancel, but some reinforce, giving bright iridescence. Light is also transmitted into, and absorbed by, the melanin granules. As with a thin film of oil on water, with different angles of viewing, the distance the light travels through the outer, transparent keratin shell varies, giving different wavelengths of light that add up and different corresponding colors—the characteristic of iridescence.

Similarly, the feathers of the shiny cowbird, European starling, and glossy ibis all derive their iridescence from a single well-defined layer of melanin granules beneath the outer shell of transparent keratin, resulting in bronze, purple, blue, or green, depending on the angle of viewing (figure 1.15). The transparent outer layer of keratin in a shiny cowbird barbule has been measured to be 0.084 microns, about one thousandth the diameter of a human hair. That feathers have such tiny structures within them producing iridescence is stunning.

FIGURE 1.14

The microscopic structure of male satin bowerbird tail feathers that produces iridescence. [(a) Alamy; (b, c, d) Doucet et al. (2006), fig. 3c,e,g. Reproduced/adapted with permission.]

(a) satin bowerbird

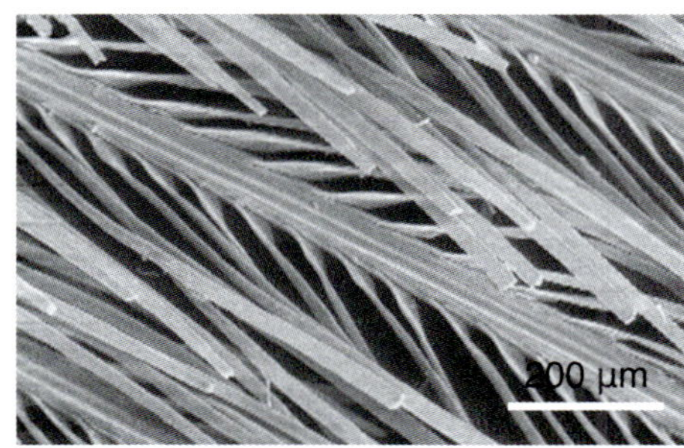

(b) barbs and barbules

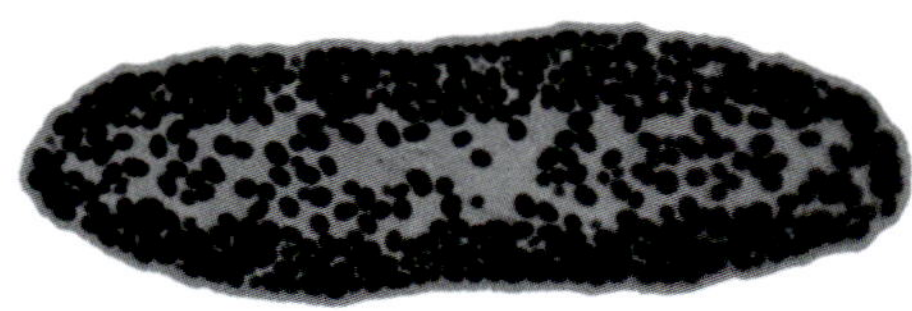

(c) barbule cross section

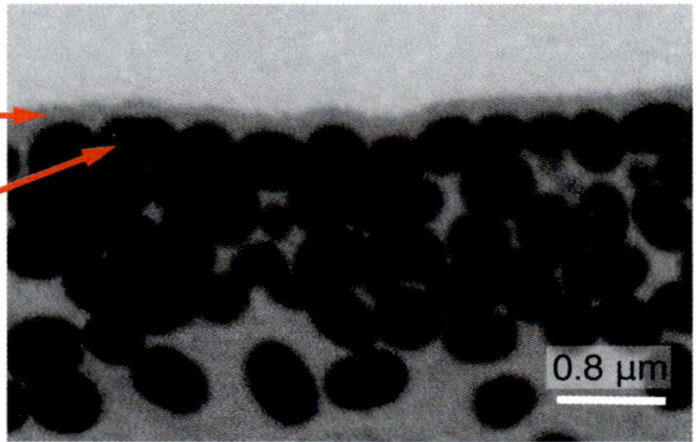

(d) cross section of barbule at higher magnification

FIGURE 1.15

(a) Shiny cowbird,
(b) glossy ibis. [Both Alamy.]

(a)

(b)

Hummingbirds achieve their spectacular iridescent colors with between 7 and 15 well-defined layers of melanosomes, pancake-shaped melanin structures with interior air pockets (figure 1.16). The multiple layers of melanosomes, with light reflected from each layer, increase the amplitude of the light wave over and over, further increasing the intensity of the color. To produce the iridescent effect, the surface of the barbules and their subsurface microstructures must be very smooth; this smoothness also gives rise to the shimmering metallic appearance of iridescent feathers.

Birds-of-paradise, which live in eastern Indonesia, Papua New Guinea, and eastern Australia, have yet another way of making their iridescent feathers striking. During courtship dances, the males of several species that are mostly black, such as Wahnes' parotia or the superb bird-of-paradise, display patches of iridescent feathers that seem to be illuminated from within the feather, or even to almost float in space, slightly away from the bird's body (figure 1.17).

Cody McCoy, of the University of Chicago (formerly of Harvard University, where she did this work), Teresa Feo of the National Museum of Natural History, and Todd Harvey and Rick Prum of Yale University have discovered how they do this. The black feathers surrounding the iridescent patches have tiny branching structures on their barbules that reflect light within the feather, so that more light is absorbed within the feather structure (rather than being reflected outward) than is the case with normal black feathers. The black feathers are "super-black," absorbing over 99% (and reflecting less than 1%) of the light that hits their surface; by comparison, normal back feathers absorb 97% (and reflect 3%) of the light that hits their surface.

Humans use the reflectance from surfaces to estimate the amount of ambient light; super-black feathers greatly reduce the reflectance, interfering with an observer's ability to estimate the amount of ambient light and the brightness of the iridescent patch. If the brain overestimates the amount of light reflecting from a patch of iridescent feathers, the feathers appear to be illuminated from within or even floating in space. McCoy and her coauthors hypothesize that the super-black feathers have the same effect on birds and that males display their iridescent patches, with the surrounding super-black feathers, to maximize this effect for females viewing their courtship dances.

The microscopic structures within feathers that give rise to structural color are astonishing. Who could have imagined that modified hairs would contain the submicron pores in spongy keratin that produce the blue of eastern bluebirds; that

FIGURE 1.16

The spectacular iridescence of hummingbirds arises from multiple layers of melanosomes. [(a) Alamy; (b) Giraldo et al. (2018), fig. 5f, CC BY 4.0.]

(a) Anna's hummingbird

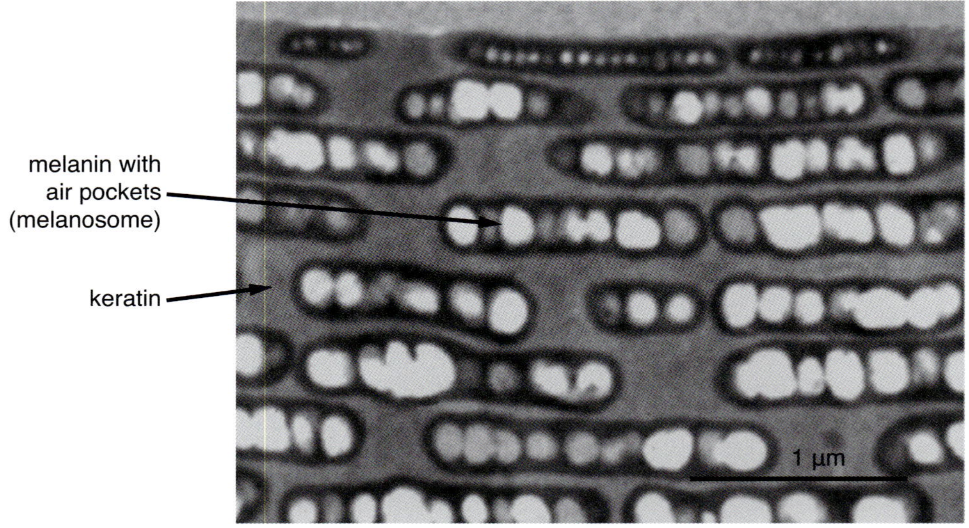

(b) Anna's hummingbird barbule cross section

(a) Wahnes' parotia

(b) superb bird-of-paradise

(c) normal black feather

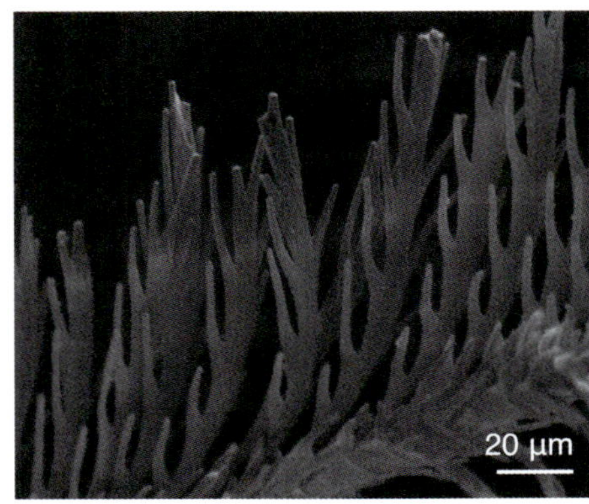

(d) highly branched barbules of a super-black feather

FIGURE 1.17

(a) Wahnes' parotia and (b) the superb bird-of-paradise both have super-black feathers, which make their iridescent patches even more dazzling. Scanning electron micrographs of (c) normal black feather showing straight barbules (from a paradise-crow) and (d) highly branched barbules of a super-black feather from Wahnes' parotia. [(a) Tim Laman; (b) Alamy; (c, d) McCoy et al. (2018), fig. 3a,b, CC BY 4.0.]

a 0.163-micron layer of keratin underlain by melanin would create the shimmering iridescence of the satin bowerbird; or that the highly branched barbules of the super-black feathers of Wahnes' parotia would make their iridescent patches appear to be lit from within?

Seeing micrographs of these tiny structures in feathers is a little like acquiring x-ray vision. In my mind's eye, whenever I see a bird with blue or iridescent feathers, I also see the microscopic structures that produce them.

FEATHERS FOR SOUND

CREATING SOUND

We saw in the last section how tiny, aligned, microscopic structures in the throat feathers of the male broad-tailed hummingbird produce iridescence. What about the trill from the male's tail feathers as he passes a female at the base of his dive? How does that work?

First, a little background on sound. If you vibrate a tuning fork, the to-and-fro motion of the tines produces small increases and decreases in the pressure of the air around it, forming a wave that travels through the air (figure 1.18). The pressure wave is detected by the ear as sound. The pitch of the sound depends on how many wavelengths reach the ear per second: the more wavelengths per second (the higher the *frequency*), the higher the pitch. A frequency of one wavelength per second is called 1 hertz (Hz), named for Heinrich Hertz, a German physicist who studied waves in the late nineteenth century. For instance, in modern tempered tuning, middle C on a piano has a frequency of 261.6 Hz, and the C one octave above that has a frequency of 523.2 Hz. (Each higher octave doubles the frequency.) Higher frequencies, in the thousands of wavelengths per second, are measured in kilohertz (kHz). The *amplitude* of a sound wave corresponds to its loudness; the greater the amplitude, the louder the sound.

It is thought, based on the frequencies at which many birds sing, that birds can hear frequencies up to at least 10 kHz. Cedar waxwings sing at frequencies between 6 kHz and 8 kHz, for example, and blackpoll warblers sing at 8 kHz to 10 kHz. By comparison, we can hear frequencies between about 20 Hz and 20 kHz—and dogs, with their more sensitive hearing, can hear frequencies of up to 60 kHz. (I recently learned that one can now buy ultrasonic squeaky toys for dogs that squeak at frequencies above 20 kHz—squeaky to dogs, thankfully silent to humans.)

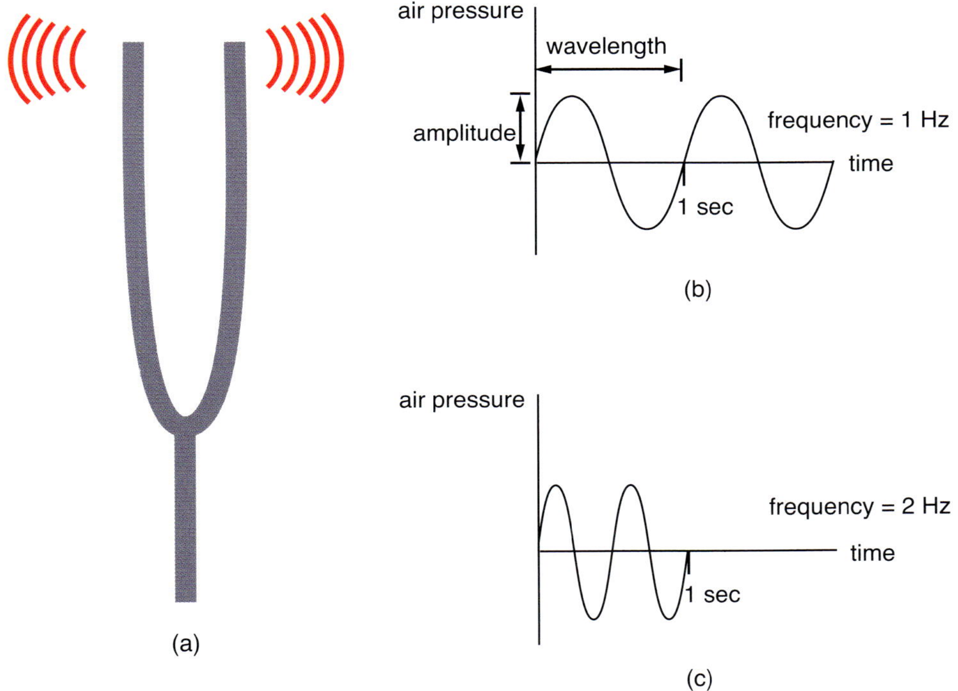

FIGURE 1.18

(a) Sound is a wave of air pressure. (b) One wavelength in one second has a frequency of 1 Hertz (Hz). (c) Two wavelengths in one second has a frequency of 2 Hz.

When a hummingbird flies by, the buzzing or humming you hear is the pressure wave created by its wings beating in flight. Ruby-throated hummingbirds beat their wings at an incredibly fast 45 times a second, creating a buzzing sound with a corresponding frequency of 45 Hertz, or a little more than two octaves below middle C on a piano. When maneuvering past obstacles, their wings flap even faster, 55–66 times a second.

So what about those broad-tailed hummingbirds? How do they produce sound with their tail feathers? Broad-tailed hummingbirds are just one species in the "bee" hummingbird group, which also includes, among others, Anna's hummingbird, rufous hummingbird, Allan's hummingbird, and an individual species called the bee hummingbird. It's not just male broad-tailed hummingbirds that do spectacular courtship dives in front of females; the males of *all* the bee hummingbird species do similar dives, making high-pitched buzzing sounds with their tail feathers at frequencies over 1 kHz at the base of their dive, in front of the female. They're desperate to impress!

When he was a graduate student at Berkeley, Christopher Clark got interested in how male Anna's hummingbirds create sound with their feathers during their courtship dives. So he and Teresa Feo, then an undergraduate, took high-speed video of male Anna's hummingbirds' dives. The video revealed that the birds started their dive with their tail feathers held close together and then, at the bottom of the dive, abruptly spread their tail feathers apart briefly, for, on average, 59 milliseconds—and it was in that interval that they made the buzzing sound.

Continuing their experimentation, Clark and Feo then manipulated the tail feathers on the birds, removing individual feathers or removing the leading or trailing vane of the outermost tail feather. In doing so, they discovered that to make the buzzing sound during a dive, the trailing vane of the outermost tail feather had to be intact. They also placed individual feathers in a jet of air: the trailing edge of the outermost tail feather fluttered at the same frequency as the buzzing sound (4 kHz). All of these observations led to the conclusion that the buzzing sound was made by fluttering of the trailing edge of the outermost tail feather.

Feather fluttering produces high-pitched sounds the same way that blowing on a blade of grass held between your thumbs produces squealing. To get a blade of grass to squeal, it has to be held with gaps between the blade of grass and the thumbs, allowing it to vibrate. And you have to blow air over the grass blade quickly, above a critical speed: if you blow over the blade too slowly, it doesn't squeal. Similarly, for a feather to flutter, a gap has to form between two feathers so that one or both are free to vibrate—by the tip of the feather bending or twisting back and forth, or by the trailing part of the vane bending up and down, or in some other way. And the speed of the air flowing over the feathers has to exceed a critical speed.

What started out as a minor side project for Clark became an obsession. As a postdoc in Rick Prum's lab at Yale, he showed, with others, that the frequency of fluttering of the tail feathers in 14 species of "bee" hummingbirds, including the broad-tailed hummingbird, matches the resonant frequency, giving the largest amplitude of vibration and loudest sound. The fluttering frequency was measured by mounting individual feathers in a wind tunnel, increasing the air speed until fluttering occurred, and measuring the shape of the feather as it vibrated by scanning it with a laser. The resonant frequency was measured by mounting individual feathers on a mini-shaker and increasing the frequency of vibration; resonance occurs when the amplitude of vibration is greatest. While there was some variation in the details of the tail feather fluttering between species (for instance, the broad-tailed hummingbird fluttered a different tail feather than the Anna's hummingbird), the buzzing sound was always created by fluttering of a tail feather.

Feather fluttering has also been observed in male African broadbills and rufous-sided broadbills, both of which produce a loud sound when a gap appears between their outer primary wing feathers, allowing them to flutter. Once again, this "wing song" appears to be part of a courtship display, similar to vocal song in songbirds; females do not display. The common snipe, too, produces a buzzing sound with its outer tail feathers during flying dives in courtship display; these are also thought to be the result of fluttering of the feathers.

Club-winged manakins—sparrow-sized birds that live in the cloud forest on the western slopes of the Andes in the north of Ecuador and southwestern Columbia—have yet another way of producing sound with their feathers (figure 1.19). A male club-winged manakin perches on a branch, flicks its wings above his back, and knocks them together, vibrating them to produce a high-pitched buzzing sound. Kim Bostwick, while a graduate student in Rick Prum's lab, took high-speed video of this behavior and found that the wings knock together an amazing 107 times each second, or at 107 Hz. But, puzzlingly, the frequency of the high-pitched buzzing sound was not 107 Hz but 1498 Hz—exactly 14 times the frequency of the wings knocking together. Something else must be happening 14 times a second to give the high-pitched sound. But what?

FIGURE 1.19

(a) Male club-winged manakin beating its wings together to make a high-pitched buzzing sound. (b) The secondary feathers. (c) The bent fifth secondary acts like a pick rubbing over (d) the ridges of the sixth secondary. [(a) Alamy; (b, c, d) Bostwick et al. (2010), fig. 1, with permission.]

(a) male club-winged manakin

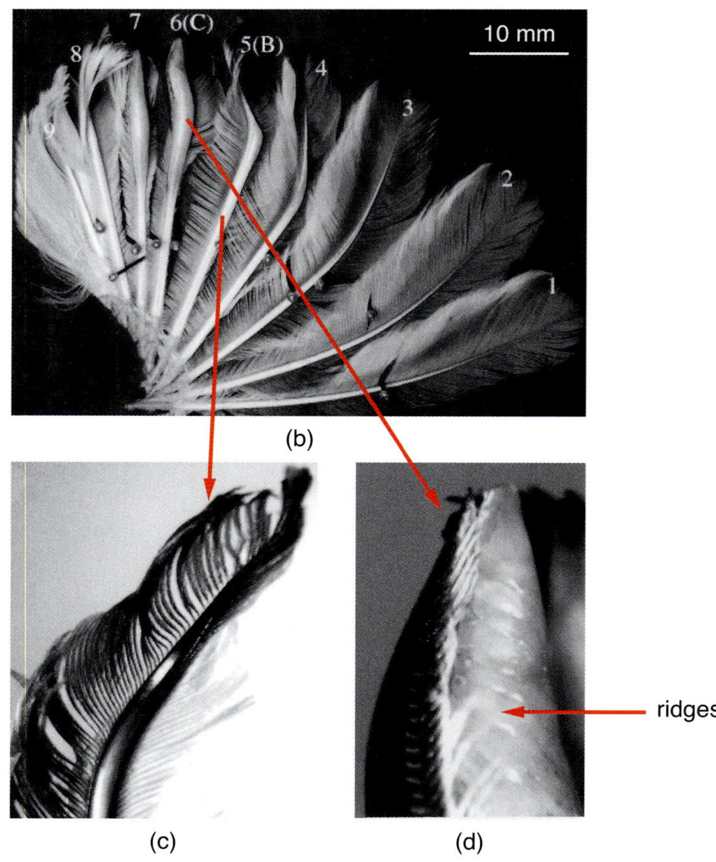

To find out, Bostwick decided to study both the primary and secondary feathers on the wing of the male club-winged manakin. (The primary feathers are the outer flight feathers along the trailing edge of the wing, attached to the "hand" bones, the manus, and the secondary feathers are inner flight feathers along the trailing edge of the wing, attached to the ulna.) When she looked closely at the secondary feathers, she noticed something very unusual: the last quarter of the fifth secondary is bent at 45° and the shafts of the sixth and seventh secondaries are unusually thick at the ends (which is where the name club-winged manakin comes from) (figure 1.19b, c, d). Looking even more closely at the tip end of the shaft of the sixth secondary, she noticed that there are 7 ridges on it, like knuckles. Bostwick realized that the bent tip of the fifth secondary lies over the sixth secondary; as the wings knock together, the tip of the fifth secondary rubs back and forth over the 7 ridges of the shaft of the sixth secondary, like a pick over a washboard, vibrating 14 times for each knocking together of the wings, producing the high-pitched buzzing sound at a frequency of 1498 Hz. The club-winged manakin's specialized secondary flight feathers act like a musical instrument. Amazing.

SUPPRESSING SOUND

Remarkably, feathers can also suppress sound. Owls, famously, fly nearly silently, thanks to microscopic structures unique to their primary and secondary flight feathers. How do they do this?

Most studies of the silent flight of owls focus on barn owls. Their primary and secondary feathers and their greater coverts (the feathers next to the primaries and secondaries) have three modifications that reduce sounds from flight (figure 1.20): serrations on the leading edge of the wing (that is, of the outermost primary and greater primary covert); fluffy fringes at the edges of each feather (except the leading edge of the outermost primary and greater primary covert); and a velvety top surface. For comparison, these features are absent in pigeon flight feathers.

The way in which these features reduce noise during flight is only partially understood. At the low flight speeds typical of barn owls, the cambered end of the wing, toward the wing tip, normally would produce separation of the airflow from the wing surface, creating noise. The serrations on the feathers of the leading edge of the wing are thought to produce a thin layer of turbulence across the width of the wing, preventing separation of the airflow from the wing surface, decreasing noise and increasing lift.

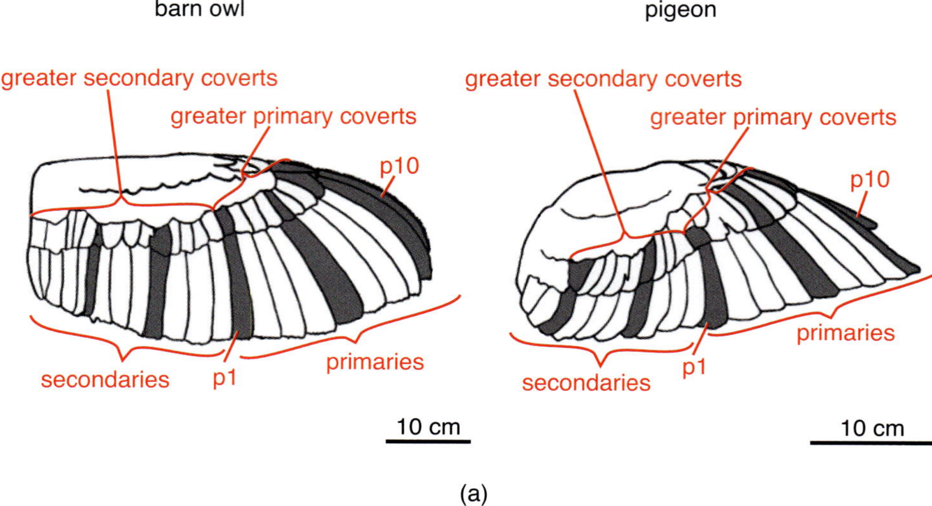

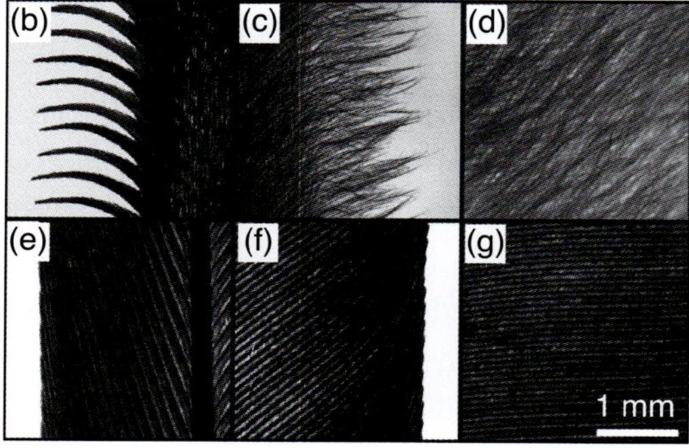

FIGURE 1.20

(a) Wing feather groups for a barn owl and a pigeon; (b, c, d) features of barn owl wing feathers that are absent in pigeon wing feathers (e, f, g). (The darker gray feathers in the top drawing are ones that Bachmann et al. used in their study.) [From Bachmann et al. (2007), fig. 1a and 7, CC BY 2.0.]

The edges of all the other primary and secondary feathers and their greater coverts have fringes. At the outer edges of the vane, the barbules lack hooklets, creating the somewhat fluffy structure of the fringe; this arrangement allows the barb to move easily. In wind tunnel tests, the fringes glide into the grooves between the parallel barb shafts of the adjacent vane, like two combs intertwining, forming a continuous wing surface. The pressure difference between the upper and lower surface of the wing during flight sucks the fringes into the adjacent feather vane, reinforcing this effect. This prevents a gap forming between primary feathers, preventing fluttering and its associated sounds.

Over the rest of the surface of the primary and secondary feathers and of their greater coverts the barbules are highly elongated and angle upward, creating the velvety top surface of the wing (figure 1.21). The barbules are so long that they cover up to four neighboring barb shafts; for comparison, the barbules in the corresponding feathers in pigeons do not reach even the adjacent barb shaft. It is thought that the elongated and upwardly angled barbules may play two roles: first, to reduce friction (and any noise associated with it) between the barbs in areas where two adjacent feathers overlap; and second, to maintain a uniform, stable airflow over the wing, reducing turbulence at the trailing edge of the wing and increasing lift.

Engineers have taken notice of the barn owl's silent flight. Most of the swooshing noise generated by wind turbine blades as they rotate is from turbulence as the air passes over and leaves the trailing edge of the blade. Siemens has introduced low-noise wind turbine blades that have a trailing edge with "comb teeth" that reduce trailing-edge turbulence; the teeth mimic the fringes at the trailing edge of owl flight feathers (figure 1.22). Others are experimenting with materials inspired by the velvety surface of owls' flight feathers. In wind tunnel experiments with air foils in the shape of a barn owl wing, air flows more uniformly over the surface, with less turbulence, if the air foil is covered with a velvety material mimicking the surface of barn owl flight feathers than if it is smooth.

Interestingly, in owls that eat fish and hunt by sight, such as Blakiston's fish-owl of northeast Asia (figure 1.23) and Pel's fishing-owl of central and southern Africa, the flight feathers lack the features that give rise to nearly silent flight; since fish are underwater and can't hear an owl approaching, there is no advantage for a fish-owl to fly silently. In flight, the wings of the brown fish-owl have even been reported making a singing noise.

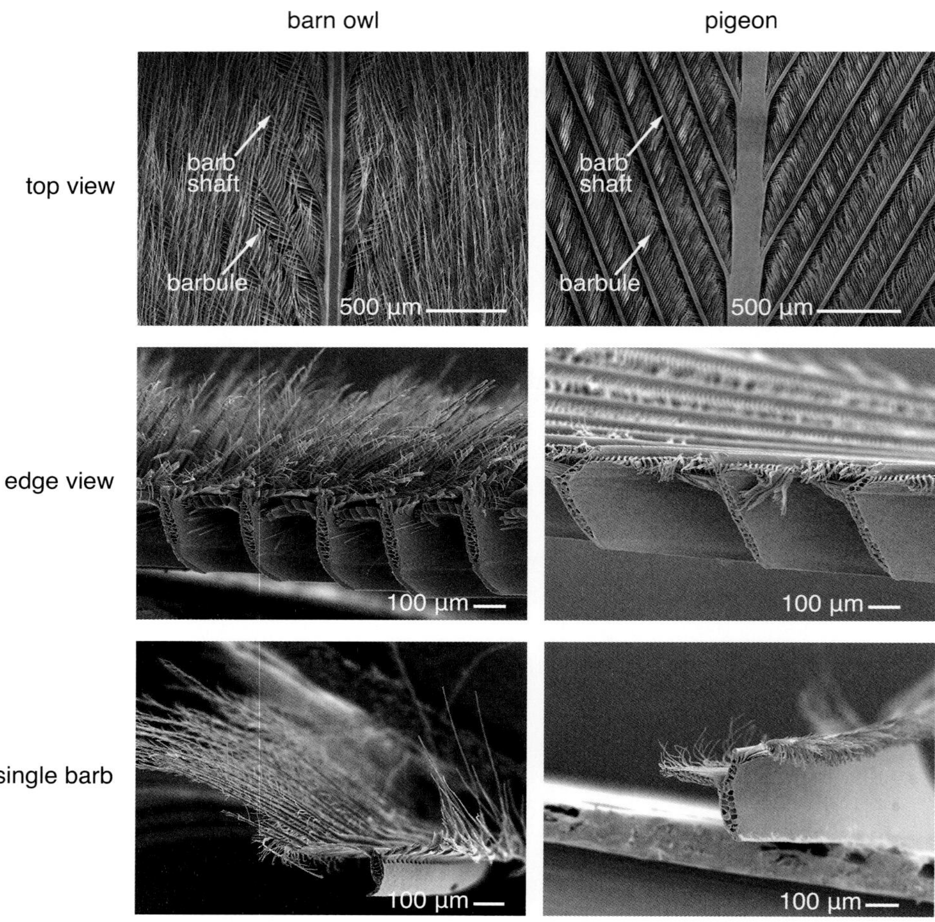

FIGURE 1.21

Left column: scanning electron microscope images of the elongated barbules of a barn owl primary feather, producing the "velvety surface." Right column: the pigeon primary feather has much shorter barbules and no velvety surface. [Scanning electron microscope images taken by Dr. Isaac Cabrera; specimens from the Massachusetts Audubon Society.]

FIGURE 1.22

A low-noise wind turbine blade with "comb teeth" that mimic the fringes on the trailing edge of a barn owl flight feather. [Alamy.]

FIGURE 1.23

Blakiston's fish-owl. Fish-owls lack the feather features that give rise to silent flight in other types of owls. [Alamy.]

COLLECTING SOUND

The feathers in the facial ruffs of owls direct sound into their ear openings, allowing them to hunt by sound alone and capture prey in total darkness. Owls can even capture prey beneath snow, sometimes leaving beautiful wing prints on the surface of the snow.

A case in point: A hungry barn owl, unfed for several days, perches in a completely darkened room at Drumlin Farm, a sanctuary run by the Massachusetts Audubon Society. A deer mouse is released into the room, rustling among the two inches of dry leaves that have been placed on the floor. As the mouse pauses, silent, the owl strikes, capturing the mouse. Over the next few days, the owl makes 16 more strikes at mice, missing only four times, each time by less than two inches.

To eliminate the possibility that the owl was using its sense of smell or detecting heat from the body of the mouse (for instance, by sensing infrared light emitted by a warm body), the experiment was repeated with a mouse-sized wad of paper dragged through the leaves; the owl again successfully hit the wad of paper. And when the owl's hearing was altered by putting cotton in one ear, the owl flew toward the mouse but missed it by about 18 inches, or a little less than half a meter.

This first demonstration that barn owls can successfully hunt in total darkness was carried out by Roger Payne, while he was an undergraduate at Harvard in the late 1950s, under the supervision of William Drury. After graduating from Harvard, Payne did further research on barn owl hearing for his PhD at Cornell University. He then took up a faculty position at Rockefeller University where, around 1967, he became interested in whale conservation. He is most famous for discovering, along with Scott McVay, the songs of humpback whales, which he described as a chorus of "exuberant, uninterrupted rivers of sound."

Barn owls detect the location of a sound in two ways: by the time it takes for a sound to reach each ear, and by the loudness or intensity of the sound in each ear. If a sound is directly ahead of the owl's face, for instance, it reaches each ear at the same time and is equally loud in each ear. But if it comes from one side of the face, it takes less time for the sound to travel to the ear on that side than the other and is louder on that side. By detecting these differences, owls can place a sound in a horizontal plane, to the left or right of the center of the face.

Many species of owl have ear openings at different heights on the left and right side of the head. You can see this asymmetry in the skull of the northern saw-whet owl (figure 1.24). In the barn owl, the left ear is higher than the right.

But it is not just the position of the ears that enables their exceptional hearing. The facial feathers play an important role, too. One of the most identifiable

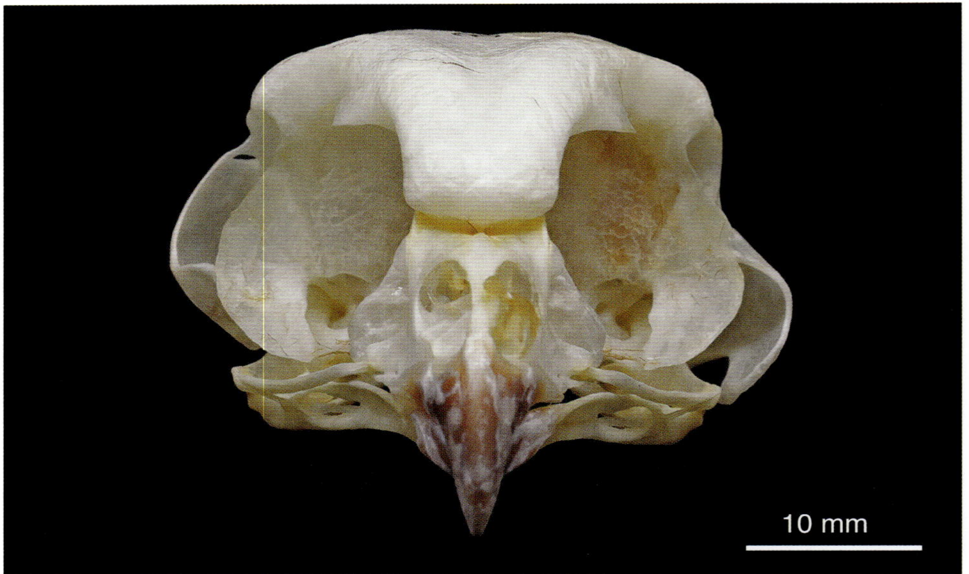

FIGURE 1.24

Northern saw-whet owl skull showing the asymmetry associated with the ears being at different heights. [Museum of Comparative Zoology, Harvard University, specimen 337061, © President and Fellows of Harvard College.]

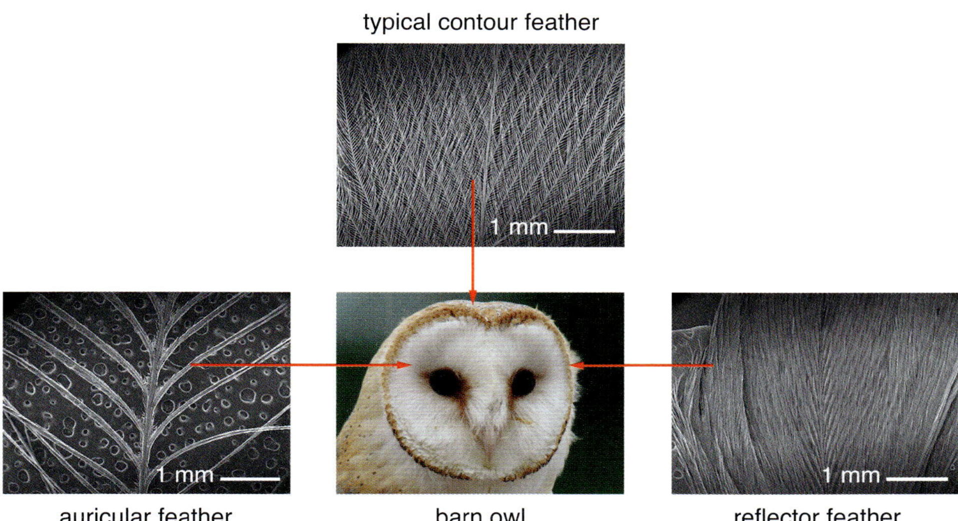

FIGURE 1.25

Barn owl auricular feathers, with their widely spaced barbs, transmit sound easily to the ruff, while the reflector feathers, with their densely packed barbs, reflect sound into the ears. (The small circular blotches below the auricular feather are an artifact of the imaging, from the electron beam melting the adhesive bonding the feather to the specimen holder.) [Barn owl image: Alamy; scanning electron microscope images taken by Dr. Isaac Cabrera. Barn owl feathers from Museum of Comparative Zoology, Harvard University, specimen 300723, © President and Fellows of Harvard College.]

features of a barn owl is its heart-shaped facial disc (figure 1.25). The disc has two specialized types of feathers that assist in pinpointing the exact location of the source of a sound: white auricular feathers, which fill the interior of the disc, and rust-colored reflector feathers, which form the ruff around the edge of the disc. The auricular feathers have fewer barbs than a typical contour feather, transmitting sound more easily through them, while the reflector feathers have more barbs than usual, reflecting sound (as their name suggests). The position of the reflector feathers focuses sound into the ear openings just within the edge of the ruff. (Fish-owls, preying on fish near the surface of rivers, lakes, and seacoasts, hunt by sight not sound; their ruffs are poorly developed [figure 1.23].)

The feathers of the ruff, too, are asymmetric: the left side of the ruff is more downward-facing than the right side, so that it preferentially focuses sound from below to the left ear, while the right side of the ruff is more upward-facing than the left, preferentially focusing sound from above to the right ear. Sounds from below are louder in the left ear while sounds from above are louder in the right ear.

Andrew Moiseff, at the University of Connecticut, demonstrated this by fitting barn owls with miniature earphones. When the sound was louder in the left ear, they turned their heads downward, but when it was louder in the right ear, the owls turned their heads upward. And in experiments in which the ruff reflector feathers were removed, done by Eric Knudsen and Masakazu Konishi at CalTech, the owls had difficulty in identifying the height at which a sound originated. The barn owl's facial disc, and especially the reflector feathers around the edge of the disc, are essential for it to be able to precisely locate the source of a sound, such as that of a mouse rustling among dead leaves.

Depending on their structure, feathers can create, suppress, or collect sound. How structure gives rise to properties is the bread and butter of materials science. Next, we'll see how the microscopic structures of a bird's inner down feathers and outer contour feathers keep them warm and dry.

REFERENCES

COLOR

Attenborough D (n.d.). Interview with Gareth Huw Davies about the BBC series *Life of Birds*. https://www.pbs.org/lifeofbirds/sirdavid/index.html

Bostwick K (2016). Feathers and plumages. In *The Cornell Lab of Ornithology Handbook of Bird Biology*, third edition, edited by Lovette IJ and Fitzpatrick JW. Wiley.

Burns KJ and Shultz AJ (2012). Widespread cryptic dichromatism and ultraviolet reflectance in the largest radiation of neotropical songbirds: implications of accounting for avian vision in the study of plumage evolution. *Auk* **129**, 211–221.

Doucet SM, Shawkey MD, Hill GE, and Montgomerie R (2006). Iridescent plumage in satin bowerbirds: structure, mechanisms and nanostructural predictors of individual variation in colour. *J. Exp. Biol.* **209**, 380–390.

Gill FB and Prum RO with contributions by Robinson SK (2019). *Ornithology*. Fourth edition. WH Freeman.

Giraldo MA, Parra JL, and Stavenga DG (2018). Iridescent colouration of male Anna's hummingbird (*Calypte anna*) caused by multilayered barbules. *J. Comp. Physiol.* **A204**, 965–975.

Hart NS, Partridge JC, Bennett ATD, and Cuthill IC (2000). Visual pigments, cone oil droplets and ocular media in four species of estrildid finch. *J. Comp. Physiol.* **A186**, 681–694.

Hill GE (2010). *Bird Coloration*. National Geographic. Chapters 4 (Pigmentary coloration) and 5 (Structural coloration)

Hogan BG and Stoddard MC (2018). Synchronization of speed, sound and iridescent color in a hummingbird aerial courtship dive. *Nature Communications* **9**, article 5260.

Hooke R (1665). *Micrographia*. Royal Society of London.

Lucas AM and Stettenheim PR (1972). *Avian Anatomy: Integument*. Agricultural Handbook 362. US Department of Agriculture.

McCoy DE, Feo T, Harvey TA, and Prum RO (2018). Structural absorption by barbule microstructures of super black bird of paradise feathers. *Nature Communications* **9**, article 1, pages 1–8.

Prum RO (2006). Anatomy, physics and evolution of structural colors. Chapter 7 in *Bird Coloration*, vol. 1, *Mechanics and Measurements*, edited by Hill GE and McGraw KJ. Harvard University Press.

Prum RO, Andersson S, and Torres RH (2003). Coherent scattering of ultraviolet light by avian feather barbs. *Auk* **120**, 163–170.

Prum RO, Dufresne ER, Quinn T, and Waters K (2009). Development of colour-producing β-keratin nanostructures in avian feather barbs. *J. Roy. Soc. Interface* **6**, S253–265.

Prum RO, Torres R, Williamson S, and Dyck J (1999). Two-dimensional Fourier analysis of the spongy medullary keratin of structurally coloured feather barbs. *Proc. Roy. Soc.* **B266**, 13–22.

Shultz A (n.d.) YouTube video *Bird Podcast*, episode 48: Behind the scenes with Allison Shultz of the Natural History Museum of Los Angeles. https://www.youtube.com/watch?v=oiqmhVEcdVc

Stoddard MC (2020). *Science Friday* interview, June 19, 2020, Hummingbirds see beyond the rainbow. https://www.sciencefriday.com/person/mary-caswell-cassie-stoddard/

CREATING SOUND

Bostwick KS (2013). Singing Wings videos, Cornell Lab of Ornithology, Bird Academy https://academy.allaboutbirds.org/singing-wings-sounds/

Bostwick KS, Elias DO, Mason A, and Montealegre-Z F (2010). Resonating feathers produce courtship song. *Proc. Roy. Soc.* **B277**, 835–841.

Bostwick KS and Prum RO (2005). Courting bird sings with stridulating wing feathers. *Science* **309**, 736.

Clark CJ (2008). Fluttering wing feathers produce flight sounds of male streamertail hummingbirds. *Biology Letters* **4**, 341–344.

Clark CJ, Elias DO, and Prum RO (2011). Aeroelastic flutter produces hummingbird feather songs. *Science* **333**, 1430–1433.

Clark CJ, Elias DO, and Prum RO (2013a). Hummingbird feather sounds are produced by aeroelastic flutter, not vortex-induced vibration. *J. Exp. Biol.* **216**, 3395–3403.

Clark CJ, Elias DO, and Prum RO (2013b). Structural resonance and mode of flutter of hummingbird tail feathers. *J. Exp. Biol.* **216**, 3404–3413.

Clark CJ and Feo TJ (2008). The Anna's hummingbird chirps with its tail: a new mechanism of sonation in birds. *Proc. R. Soc. Lond. B.* **275**, 955–962.

Clark CJ, Kirschel ANG, Hadjioannou L, and Prum RO (2016). *Smithornis* broadbills produce loud wing song by aeroelastic flutter of medial primary wing feathers. *J. Exp. Biol.* **219**, 1069–1075.

Clark CJ and Prum RO (2015). Aeroelastic flutter of feathers, flight and the evolution of non-vocal communication in birds. *J. Exp. Biol.* **218**, 3520–3527.

Cornell All About Birds (n.d.). https://www.allaboutbirds.org/do-bird-songs-have-frequencies-higher-than-humans-can-hear/

Gill FB and Prum RO, with contributions by Robinson SK (2019). *Ornithology*. Fourth edition. WH Freeman.

Hunter TA and Picman J (2005). Characteristics of the wing sounds of four hummingbird species that breed in Canada. *Condor* **107**, 570–582.

van Casteren A, Codd JR, Gardiner JD, McGhie H, and Ennos AR (2010). Sonation in the male common snipe (*Capella gallinago gallinago* L) is achieved by a flag-like fluttering of their tail feathers and consequent vortex shedding. *J. Exp. Biol.* **213**, 1602–1608.

SUPPRESSING SOUND

Bachmann T, Blazek S, Erlinghagen T, Baumgartner W, and Wagner H (2012). Barn owl flight. In *Nature-Inspired Fluid Mechanics*, edited by Tropea C and Bleckmann H, 101–117. Springer Verlag.

Bachmann T, Klan S, Baumgartner W, Klaas M, Schroder W, and Wagner H (2007). Morphometric characterisation of wing feathers of the barn owl *Tyto alba pratincola* and the pigeon *Columba livia*. *Frontiers in Zoology* **4**, 23–38.

Bachmann T and Wagner H (2011). The three-dimensional shape of serrations at barn owl wings: towards a typical natural serration as a role model for biomimetic applications. *J. Anatomy* **219**, 192–202.

Bachmann T, Wagner H, and Tropea C (2012). Inner vane fringes of barn owl feathers reconsidered: morphometric data and functional aspects. *J. Anatomy* **221**, 1–8.

BBC barn owl flight video: https://www.youtube.com/watch?v=d_FEaFgJyfA

Clark IA, Daly CA, Devenport W, Alexander WN, Peake N, Jaworski JW, and Glegg S (2016). Bio-inspired canopies for the reduction of roughness noise. *J. Sound and Vibration* **385**, 33–54.

Klan S, Burgmann S, Bachmann T, Klaas M, Wagner H, and Schroder W (2012). Surface structure and dimensional effects on the aerodynamics of an owl-based wing model. *Eur. J. Mechanics B Fluids* **33**, 58–73.

Lederer RJ (2016). *Beaks, Bones and Bird Songs: How the Struggle for Survival Has Shaped Birds and Their Behavior*. Timber Press.

Siemens (2016). Press release. https://www.siemens.com/press/en/pressrelease/?press=/en/press-release/2016/windpower-renewables/pr2016090418wpen.htm

Voous KH (1989). *Owls of the Northern Hemisphere*. MIT Press.

Wagner H, Weger M, Klaas M, and Schroder W (2016). Features of owl wings that promote silent flight. *Interface Focus* 7, 2016.0078.

COLLECTING SOUND

Gill FB (2007). *Ornithology*. Third edition. WH Freeman.

Knudsen EI and Konishi M (1979). Mechanisms of sound localization in the Barn Owl (*Tyto alba*). *J. Comp. Physiol.* **133**, 13–21.

Moiseff A (1989). Bi-coordinate sound localization by the barn owl. *J. Comp. Physiol.* **A164**, 637–644.

Payne RS (1971). Acoustic location of prey by barn owls (*Tyto alba*). *J. Exp. Biol.* **54**, 535–573.

Payne RS and Drury WH (1958). Marksman of the darkness (*Tyto alba*). *Natural History* **67**, 316–323.

Roberts S (2023). Roger Payne, biologist who heard whales singing, dies at 88. *New York Times*, June 15.

von Campenhausen M and Wagner H (2006). Influence of the facial ruff on the sound-receiving characteristics of the barn owl's ears. *J. Comp. Physiol.* **A192**, 1073–1082.

Voous KH (1989). *Owls of the Northern Hemisphere*. MIT Press.

2

FANTASTIC FEATHERS: WARM AND DRY

The wind whipping through the northern spruce-fir forest sounds like pounding surf, even as the thermometer routinely reads −20°C [−4°F]—and sometimes −30°C [−22°F]. . . . The cold is no mere abstraction. How do kinglets, who are out there day and night and who are no bigger than the end of my thumb, maintain their body temperature of 43° to 44°C [109° to 111°F]? . . . I feel wonder, and . . . am still awestruck that anything as small and as dependent on keeping warm can survive.
Bernd Heinrich

For me, winter means birding along the shore of Cape Ann, about an hour north of Boston, to see ducks. The drive along the shoreline is gorgeous, with views of rocky promontories, small bays, and islands offshore. But the main attraction is the birds. Looking out over Gloucester Harbor, I see red-breasted mergansers, buffleheads, and a couple of rafts of common eiders. Further along, a small group of common goldeneyes, close enough to get a good look at their rather stunningly yellow eyes, which contrast with their black to greenish heads. White-winged and surf scoters. Red-throated and common loons. Best of all, a couple of absurdly colorful harlequin ducks, with their slatey-blue bodies trimmed in bands of black, white, and rust, at a huge granite pier jutting out into the water, and another half a dozen close to the shore at Halibut Point, at the northern tip of Cape Ann.

It's cold, really cold. And windy. I'm bundled up in my long underwear, jeans, flannel shirt, wool sweater, down parka, wool hat, and gloves, loving seeing the birds but also thankful every time I get back into the car and can warm up. The ducks are out on the water, all day and night, diving for fish and invertebrates. How do they stay warm?

The short answer is that they do it with their down, which traps air under their contour feathers, creating a layer of insulation that keeps body heat in. But down can only trap air if it's dry, so to figure out how birds stay warm, we first have to figure out how they stay dry.

WATER REPELLENCY

Water off a duck's back: there it is, sheets of it, running off the mallard in figure 2.1. What makes those feathers so water-repellent?

It's well known that birds preen their feathers using oils and waxes from the preen gland (the uropygial gland) near the base of their tails (figure 2.2). For many years, ornithologists thought that it was these oils and waxes that made feathers water-repellent. But feathers maintain their water repellency even when cleaned of oils and waxes, and even when the preen gland is removed when birds are hatchlings. So if the oils and waxes from the preen gland don't give feathers their water repellency, what do they do? It is now thought that they help maintain the flexibility of feathers by preventing moisture loss within the feather material, keratin, which becomes stiffer and more brittle when dried out. Our fingernails are also made of keratin; the preening of feathers is a little like applying hand lotion to one's fingernails. The oils and waxes also combat bacteria and fungi attracted to feathers. But if the products of the preen gland don't make feathers water-repellent, what does?

To be water-repellent, the surface of a feather has to cause water to bead up and run off it, rather than penetrate it. If there is a lot of water, the water beads will coalesce to form a sheet, without penetrating the feather vane, and the sheet of water will then run off the feathers (as in the photo of the mallard). Surfaces that repel water are said to be *hydrophobic* (water-fearing).

But there are degrees of water repellency. Wax, for example, is more water-repellent than glass. If you put a few drops of water on a sheet of paraffin wax, it beads up, whereas if you put it on a sheet of glass, it spreads out.

FIGURE 2.1

Sheets of water off a duck's back. [Alamy.]

FIGURE 2.2

Great egret preening. [Alamy.]

FIGURE 2.3
Contact angles for water on (a) glass, (b) paraffin wax, and (c) keratin of a Canada goose feather shaft. The feather shaft is nearly as water-repellent as wax. [Photos taken by Drs. Isaac Cabrera and Geetha Berera.]

contact angle for water on:

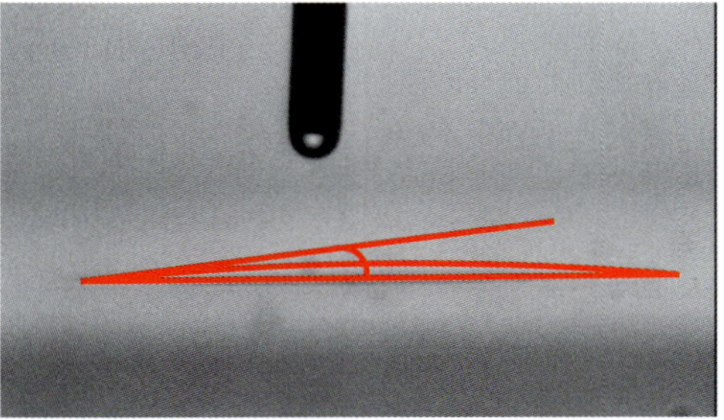

(a) glass 7°

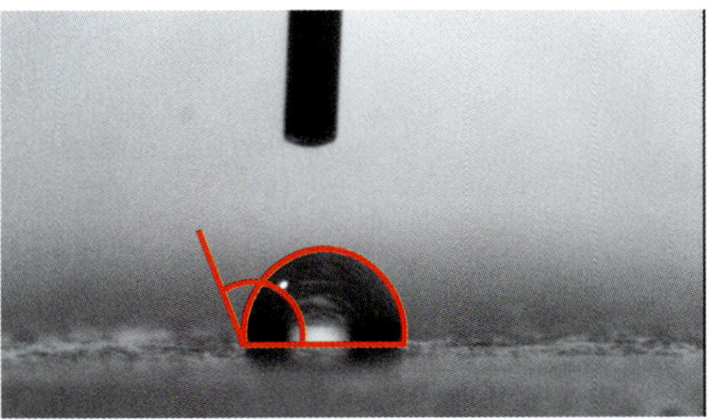

(b) paraffin wax 109°

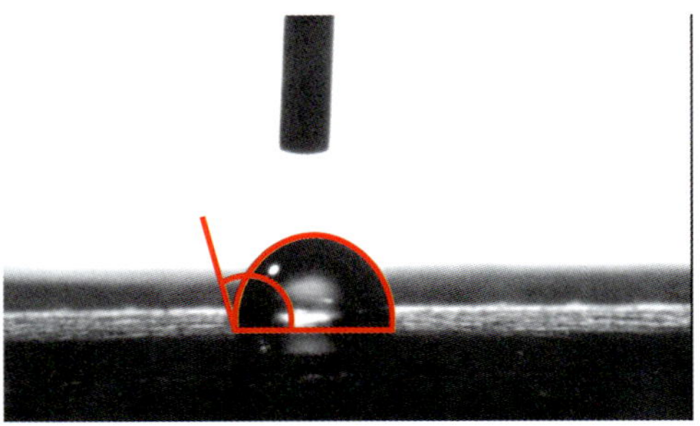

(c) keratin 102°

The spreading of a liquid on a flat solid surface is characterized by the contact angle: that is, the angle between the horizontal surface and the tangent to the liquid at its outer edge, where it contacts the air (figure 2.3). The more water is beaded up, the higher the contact angle, and the more water-repellent or hydrophobic the surface is. Measurements of contact angle, done with an instrument designed for this purpose, give values of the contact angles for water on glass of 7°, on wax of 109°, and on a feather shaft of 102°: the keratin of the feather shaft is nearly as water-repellent as wax.

Why does water bead up more on some surfaces than others? How do the surfaces of different materials differ?

Let's first see how atoms at the surface of a solid material differ from those in the interior of the solid: figure 2.4 shows a block of material (in blue) along with its surface (in red); the atoms within a small square within that block are shown on the right. In the interior of the material (the blue atoms), each atom is surrounded on all sides by other atoms of the same material. But at the surface, there are no atoms above the surface, so that the bonding of an atom at the surface differs from that of one in the interior of the solid. Atoms in the interior are more stable, and have lower energy, than atoms at the surface. The difference in energy between an atom at the surface and one in

FIGURE 2.4

Atoms in the bulk of a material are bonded to neighboring atoms in all directions. Atoms at the surface have no neighbors above them, so that the bonding at the surface differs from that in the interior.

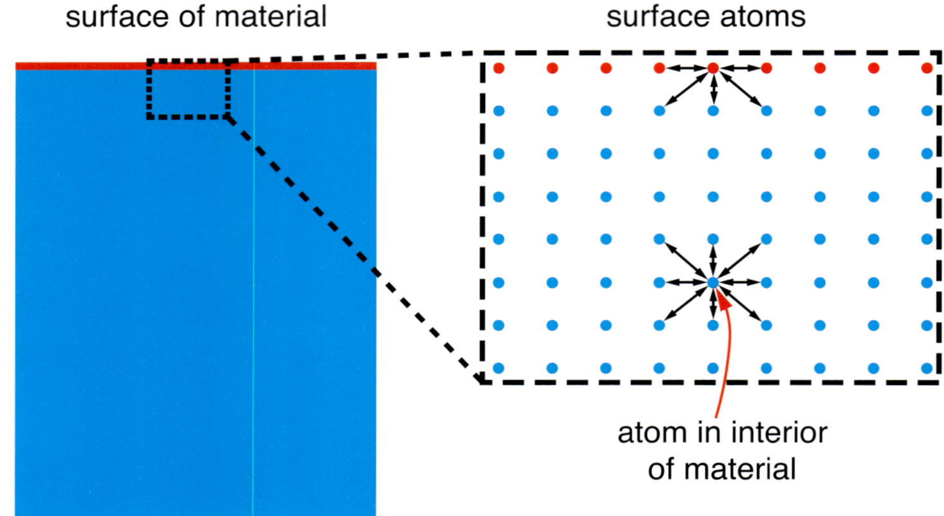

the interior is called the *surface energy*. Surface energy has units of energy per unit surface area; in the metric system, energy has units of joules, and surface energy has units of joules per square meter (joules/meter^2).

The stronger the bonding between atoms in a material, the higher the surface energy. The surface energy of clean glass is nearly 50 times that of wax (3 vs. 0.065 joules/meter^2). So one difference between glass and wax is that glass has a much, much higher surface energy. Water, too, has a surface energy: its value is a little higher than that of wax (0.072 joules/meter^2). In liquids, the surface energy is also called the surface tension. Water striders walking across the surface of a still pond are held up by surface tension. If the pond becomes polluted by detergents, which lower the surface tension, they sink and drown.

When you put a drop of water onto a solid flat surface, the drop will take up the shape that minimizes the overall energy (in joules): the most stable shape is the one that minimizes energy (figure 2.5). Water has a lower surface energy than glass, so the overall energy (in joules) is reduced by the water spreading out over the glass, increasing its surface area and reducing that of the glass. On the other hand, water has a slightly higher surface energy than wax; the overall energy (in joules) is reduced by the water beading up on the wax surface. (At the interface between two different materials, there is also an interfacial energy, associated with the difference in surface energies of the two materials, but we won't delve into that here.)

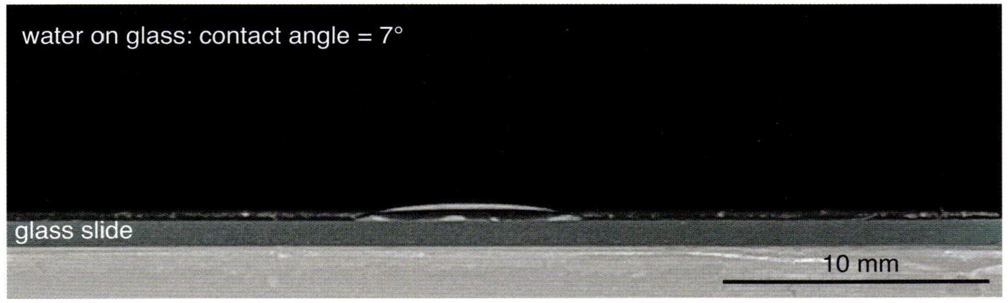

glass: 3 joules/meter^2
water: 0.072 joules/meter^2

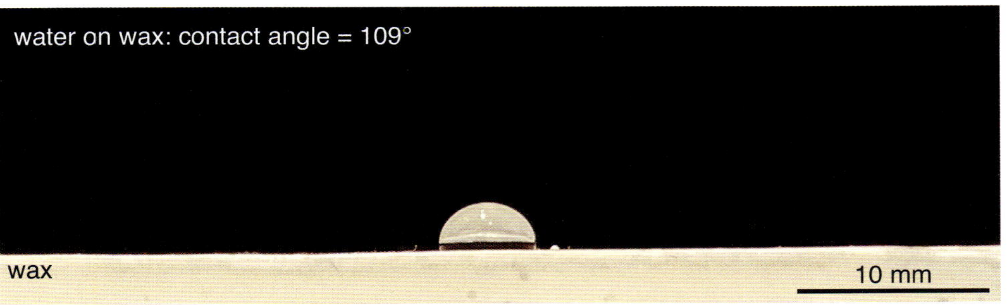

wax: 0.065 joules/meter^2
water: 0.072 joules/meter^2

FIGURE 2.5

Water takes a shape that minimizes the overall energy (in joules). (a) Glass has a higher surface energy than water, so the energy (in joules) is minimized by reducing the surface area of glass and increasing the surface area of water. (b) Water has a higher surface energy than wax, so the overall energy (in joules) is minimized by reducing the surface area of water and increasing the surface area of wax.

The more water-repellent a surface is, the more water beads up on it and the higher the contact angle. As we've seen, wax is very water-repellent, with a contact angle of 109°. And solid keratin is nearly as water-repellent as wax, with a contact angle of 102°. So part of the water repellency of feathers arises from the bonding and surface energy of solid keratin.

But this is not the entire story of how feathers repel water. Water is repelled mostly by the vane, not the solid shaft, of a contour feather. Might the structure of the barbs and barbules within the vane contribute to the water repellency of a feather?

During World War II, A. B. D. Cassie and S. Baxter, working at the Wool Industries Research Association, in Leeds, England, were interested in ways to make military clothing, such as wool greatcoats, more water-repellent. They were especially interested in how wool fibers interact with water (figure 2.6). Wool fibers intertwine with each other in a rather messy, complex way. To simplify things, the researchers instead looked at an array of identical, parallel fibers and realized that a water droplet might bridge the spacing between fibers, so long as the distance between the fibers was not too great compared with the size of the droplet. They modeled wool fibers as a series of identical, parallel, evenly spaced cylindrical rods and worked out the apparent contact angle for the array of rods, based on the surface energy and area of each surface. And what they found was quite remarkable. For rods of a constant diameter, the apparent contact angle (and hence the water repellency) can be increased dramatically by increasing the spacing between the rods.

They then came up with the idea of applying their analysis to duck feathers, with their array of parallel barbules acting as the rods in the model. From their measurements of the barbule radius (4 microns) and spacing (32 microns), they predicted the contact angle to increase dramatically, from about 100° on the solid feather shaft, to roughly 140° on the feather vane. You can see what this looks like in the photographs of water drops on a Canada goose contour feather shaft and vane in figure 2.7; the contact angle increases from 102° on the solid keratin shaft to 142° on the vane. An approximate calculation of the apparent contact angle for feathers using the Cassie-Baxter equation is given in box 2.1.

(a)

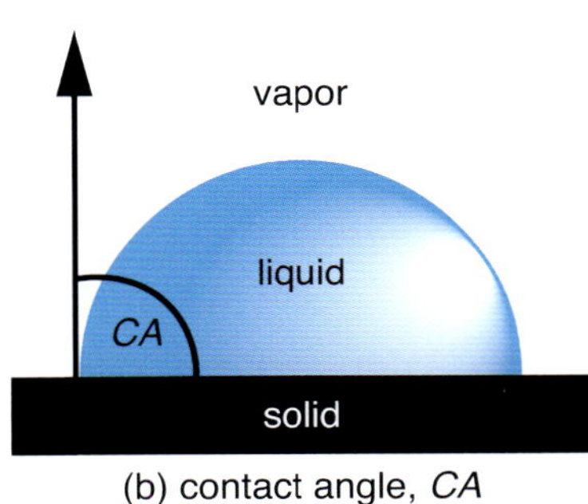

(b) contact angle, *CA*

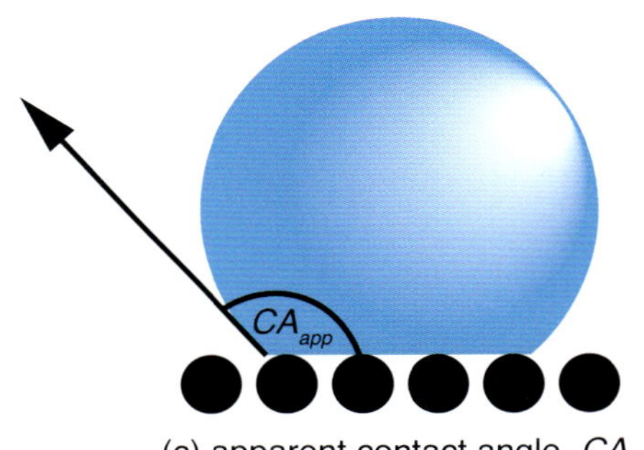

(c) apparent contact angle, CA_{app}

FIGURE 2.6

(a) British army men in their wool greatcoats. (b) Water droplet on a solid surface; (c) water droplet on identical, parallel, evenly spaced cylindrical rods. The apparent contact angle, CA_{app}, calculated by Cassie and Baxter, increases with the spacing of the rods (up to a limit). (The apparent contact angle is just the contact angle for the array of rods, rather than for the flat, solid surface.) [(a) Alamy; (b, c) after https://en.wikipedia.org/wiki/Ultrahydrophobicity#/media/File:Contact_angle_microstates.svg by Vladsinger CC BY-SA 3.0.]

FIGURE 2.7

Water on a cleaned Canada goose flight feather (a) shaft, with a contact angle of 102°, and (b) vane, with a contact angle of 142°. [Photos taken by Drs. Isaac Cabrera and Geetha Berera.]

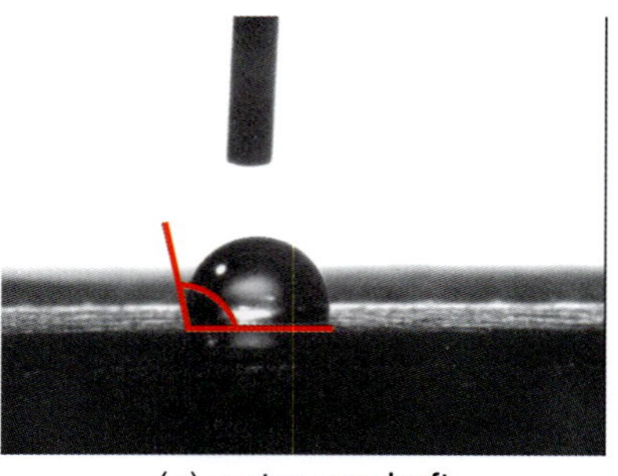

(a) water on shaft
contact angle = 102°

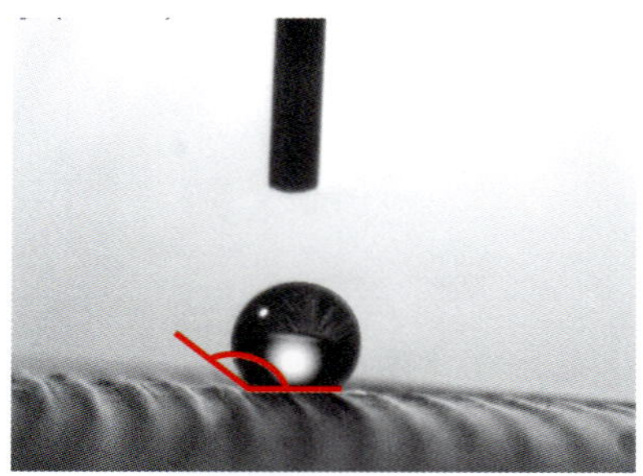

(b) water on vane
contact angle = 142°

From photographs of water droplets on a cleaned pigeon contour feather vane, Edward, Yelena, and Ester Bormashenko and their co-workers, Tamir Stein and Gene Whyman, of the College of Judea and Samaria in Israel, have measured even higher apparent contact angles, between 150° and 164° in some regions of the vane. Engineers call such surfaces "superhydrophobic."

The highest theoretically possible value for a contact angle is just under 180°, corresponding to the water droplet contacting the surface at just a tiny point. At contact angles over 140°, the area of contact between the water and the feather vane is very small so that the droplet rolls off the surface easily. Like water off a duck's back.

Cassie and Baxter conclude their classic paper with this comment: "it is sufficient to note that water rolls off a duck's back because of the structure of its feathers rather than because of any exceptional proofing agent, and that man's attempts to make clothing with the water repellency of a duck should be directed to perfecting an appropriate cloth structure rather than, as at present, to searching for an improved water-repellent agent."

Today, fabrics for raincoats sometimes use a "durable water-repellent" (DWR) outer layer that features microscopic polymer posts that project out from the surface of the fabric (figure 2.9). The posts mimic the surface structure of the lotus leaf, also well known for its water repellency. The posts in the durable water-repellent fabric are designed based on the Cassie-Baxter model, modified for the slightly different geometry of parallel vertical posts rather than the parallel barbules of a contour feather. As with the feather barbules, the geometry of the posts increases the contact angle and the water repellency.

BOX 2.1: CASSIE-BAXTER MODEL FOR FEATHERS

If the contact angle, *CA*, between water and a solid material is 90°, as it roughly is for solid keratin, Cassie and Baxter's equation for the apparent contact angle, CA_{app}, of water resting on evenly spaced, parallel rods of radius *r* of that material, separated by a distance 2*d* (figure 2.8), simplifies to:

$$\cos\left(CA_{app}\right) = \frac{r}{r+d} - 1$$

Feather barbules typically have values of $\frac{r+d}{r}$ between 2 and 8; the corresponding apparent contact angles are given in the table below.

$\frac{r+d}{r}$	Apparent contact angle, CA_{app} (°)
1	90
2	120
4	139
6	146
8	151

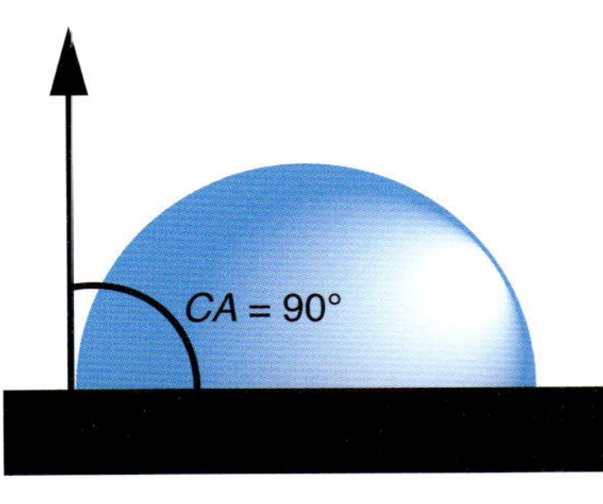

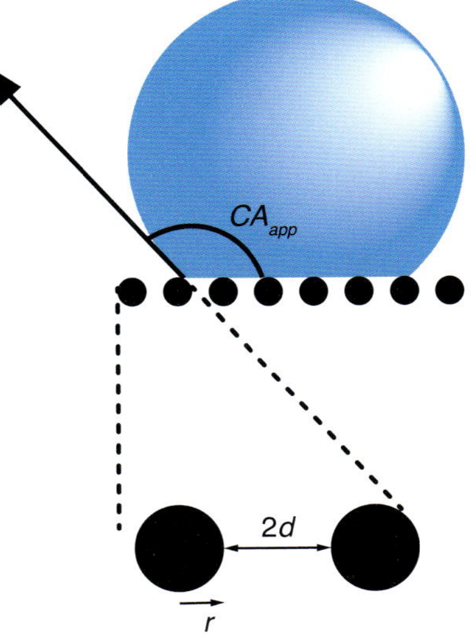

FIGURE 2.8

Cassie-Baxter model for feathers. (a) Water droplet on a solid surface; (b) water droplet on identical, parallel, evenly spaced cylindrical rods, showing radius, *r*, of a rod and the spacing, 2*d*, between rods. [After https://en.wikipedia.org/wiki/Ultrahydrophobicity#/media/File:Contact_angle_microstates.svg by Vladsinger CC BY-SA 3.0.]

FIGURE 2.9

(a) Water beading up on a durable water-repellent (DWR) fabric. (b) Schematic of the structure of a durable water-repellent fabric, showing the vertical posts that mimic the surface of a lotus leaf. [(a) Alamy.]

(a) durable water-repellent fabric

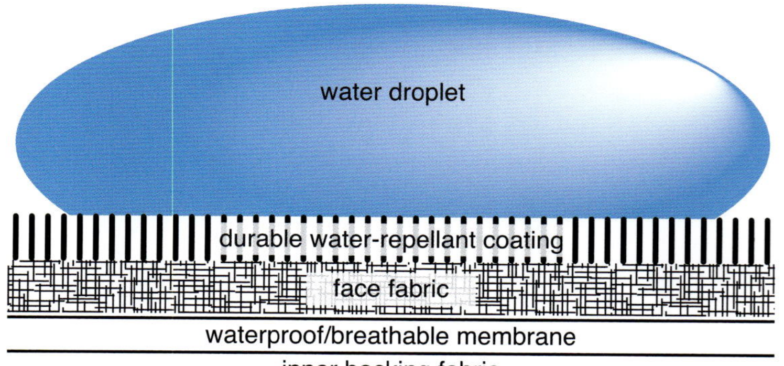

(b) schematic of the fabric structure

BOX 2.2: CORMORANTS

Cormorants are well known for their habit of standing on a rock or perching in a tree with their wings and tails spread out, drying off after a dive. During their dives, which typically are to a depth of less than 30 feet (10 m) and last less than 30 seconds, the flight feathers on their wings and tails get wet, which helps to reduce their buoyancy, making it easier for them to dive. But if the flight feathers got entirely wet, their down would lose its insulation value, and the birds would quickly become hypothermic. They solve this problem with two distinct zones in their contour feathers: an inner, water-repellent zone with the usual structure of parallel barbs that hold the vane together with interlocking hooklets and grooves, and an outer zone with a loose, irregular structure that water can penetrate (figure 2.10). During a dive, the water absorbed in the outer zone decreases buoyancy, making it easier to dive, while the inner zone maintains its water repellency, keeping the down dry to retain body heat.

FIGURE 2.10

Great cormorant contour feather, showing the inner, water-repellent layer, with the barbs spaced closely together, and the outer, water-absorbing layer, with the barbs spaced further apart. [Museum of Comparative Zoology, Harvard University, specimen 335716, © President and Fellows of Harvard College.]

THERMAL INSULATION

Some birds can survive even the most brutal polar temperatures. While some snowy owls migrate south for the winter—seeing them at Cape Ann and Plum Island, north of Boston, is magical—others remain in the Arctic, braving average January temperatures that hover around −22°F (−30°C). Some populations of common eider, too, winter along the coast of Greenland and southern Alaska. Emperor penguins breed on the ice in the Antarctic winter where temperatures drop to as low as −76°F (−60°C) and winds can reach 120 mph (200 km/hr) (figure 2.11).

Even common backyard birds are impressive in their ability to withstand cold. Black-capped chickadees don't migrate, maintaining their year-round range even in southern Alaska, and northern cardinals winter as far north as Montreal and Maine.

How do they do it?

By insulating themselves with down. When you look at a duck, for instance, you don't realize just how much of it is down. One study found that the plumage of ducks typically makes up 25–40% of their volume, and that air within the plumage—nearly all of it trapped in down—makes up about 80% of the volume of the plumage. So trapped air probably makes up something like 20–32% of a duck's volume. The remarkable buoyancy of ducks, which keeps nearly their entire bodies out of the water as they paddle, too, relies on the air trapped in down.

Emperor penguins, which tolerate the world's coldest temperatures, have a mind-boggling number of down feathers: one study estimated 108,000, which translates to one down feather every twentieth of an inch (or 1.3 mm) across the surface of the skin.

Down feathers are made up of barbs that radiate out from the end of the feather shaft, with barbules branching off the barbs (figure 2.12). Fossils of down-like feathers, with simple barbs radiating out from the end of a shaft, are among the earliest fossil feathers yet discovered; it is thought that they, too, were probably for thermal insulation.

Down insulates a bird in roughly the same way that foam insulates a building. While there are differences in the structure of down and foams (down feathers are fibrous, and foams are cellular), there are two important similarities for insulating against the cold: both are mostly air, with little solid (down on a bird is roughly 99% air by volume, and the foam in figure 2.12 is 97% air by volume), and in both the spaces between the solid are small (about 0.1 mm in the down feather and 0.25 mm in the foam).

(a) snowy owl

(b) emperor penguins

FIGURE 2.11

(a) Snowy owls live through winter in the Arctic, and (b) emperor penguins survive blizzard conditions in Antarctica. [Both Alamy.]

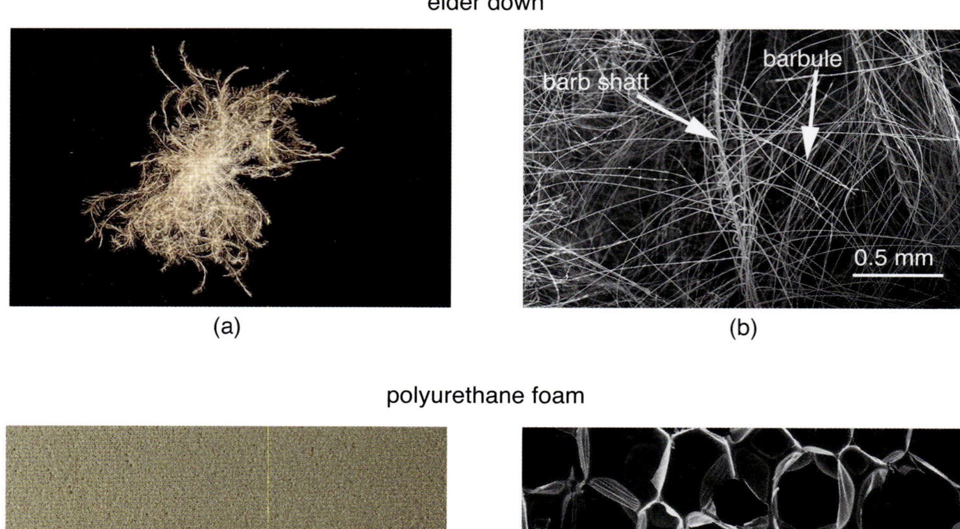

FIGURE 2.12

(a) Eider down feather, (b) scanning electron microscope image of eider down, (c) polyurethane insulation foam, (d) scanning electron microscope image of polyurethane foam. [Eider down supplied by Plumeria Bay; polyurethane foam supplied by General Plastics; scanning electron microscope images (b) taken by Dr. Isaac Cabrera and (d) from Gibson and Ashby (1997), © Cambridge University Press, Reprinted with permission.]

Outdoors, on a cool fall night with a campfire going, you can get a sense of the three ways that heat flows (figure 2.13). First, if you prod the fire with one end of an iron poker for any length of time, you will notice that the entire length of the poker heats up: heat *conducts* along the poker. The hot air above the fire rises (since hot air is less dense than cooler air), producing *convection*. And finally, if you stand in front of the fire, you can feel heat *radiating* off it. We'll next look at each of these ways that heat flows to see how the microscopic structure of down reduces heat flow to insulate a bird.

First, let's look at *conduction*, by considering that iron poker in the campfire. When you first put it in, the end of the poker in the fire is hotter than the end out of the fire. Atoms in the hot end of the poker have more energy and vibrate more than neighboring atoms that are not quite as hot. In doing so, the hotter atoms jostle neighboring cooler atoms, imparting some of their energy to them, making them hotter and eventually heating up the poker along its length.

The ability of a material to conduct heat is characterized by its *thermal conductivity*. The higher the thermal conductivity, the faster heat is conducted along a material. And the lower the thermal conductivity, the more slowly heat is conducted along a material and the more insulation the material provides. In table 2.1, the values for thermal conductivity of air and keratin, the solid material in feathers, are compared with those of polyurethane, water, and copper. Thermal conductivity has unfamiliar, rather messy, units of watts/(meter × degree kelvin), or W/mK. (The US/Imperial units are even messier and I'll avoid using them here.) The important thing here is the value relative to air, in the right-hand column.

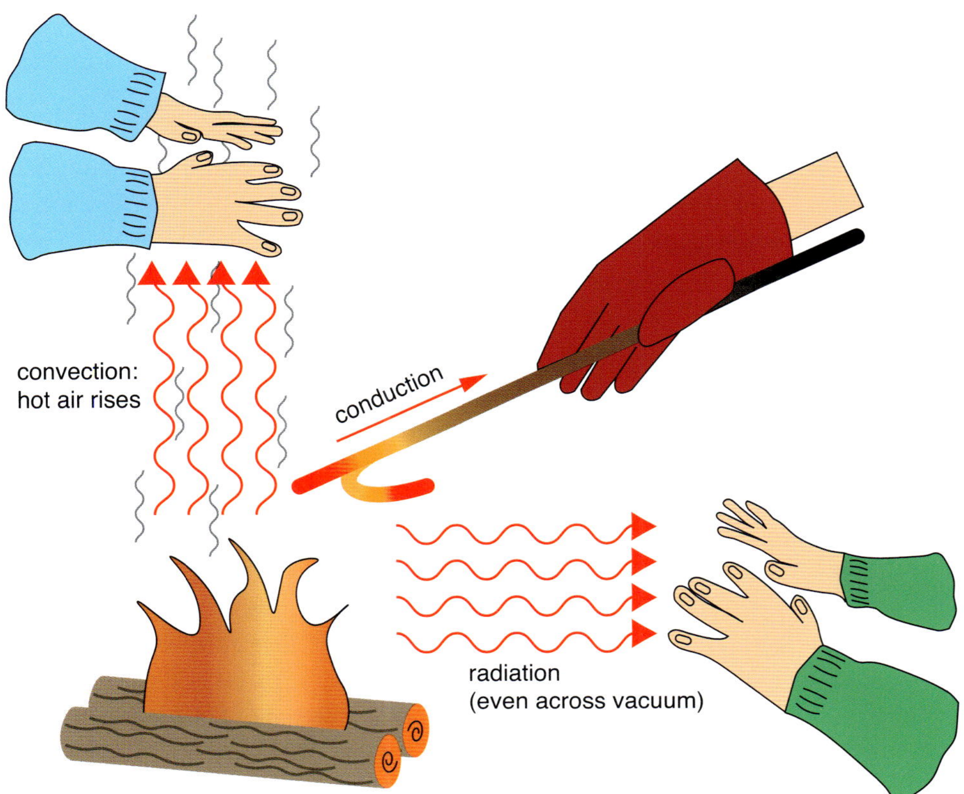

FIGURE 2.13

The three ways that heat flows: convection, conduction, and radiation. [After Kmecfiunit, cmglee, CC BY-SA 4.0, https://en.wikipedia.org/wiki/Heat_transfer.]

Table 2.1: Thermal conductivities of materials

Material	Thermal conductivity (W/mK)	Thermal conductivity relative to air
Air	0.025	1
Keratin (feather material)	0.20	8
Polyurethane	0.25	10
Water	0.60	24
Copper	400	16,000

Data from Chemical Rubber Company (1985); Granta Design (n.d.); Schuetz and Glicksman (1983); http://www.engineeringtoolbox.com/thermal-conductivity-d_429.html.

Of the materials listed, air has the lowest thermal conductivity and so is the best insulator against heat loss. The low thermal conductivity of air arises from its low density compared to solids. There are far fewer atoms in a given volume of a gas than a solid, so there's less chance of atoms jostling one another and transferring energy in a gas. Because of this, gases have much lower thermal conductivities than liquids or solids.

Keratin, the material that feathers are made of, and polyurethane, the material of our insulating foam, have similar values of thermal conductivity, about 8–10 times that of air. Water conducts heat away from the body over 20 times faster than air: this is the reason that we feel colder when wet, and that down, either on a bird or in a sleeping bag, must be kept dry to insulate well. The conductivity of copper, one of the best conductors of heat, is 16,000 times that of air. A copper pipe at room temperature feels cold to touch because it quickly conducts heat away from your hand, making your hand feel cold.

In a porous material, such as down feathers, heat flows by conduction through both the solid and the gas and so depends on the relative amount of each. Down feathers are excellent insulators largely due to their high fraction (about 99%) of low-conductivity air and the very small fraction (about 1%) of higher-conductivity keratin (figure 2.14).

Heat also flows by *convection*. In a room with a fire going in a woodstove, hot air around the stove rises, and cooler, denser air falls, creating a convection current within the room. But in foam, with its small pores, typically less than a few millimeters, friction from the motion of the air past the surface of the foam cell walls completely suppresses convection. The same thing applies in down feathers: the small spacing between the barbs and barbules suppresses convection.

Finally, heat flows by *radiation*, which occurs when electromagnetic waves are emitted from a surface (as from a woodstove, for instance, or an infrared lamp that keeps food warm in a restaurant kitchen): the hotter the surface, the more heat is emitted from it by radiation. Radiation occurs even across a vacuum; the sun's energy is transmitted across space by radiation.

If foam is placed between two plates at different temperatures, radiant heat will be absorbed and reflected at each cell wall within the foam. The smaller the pore size in the foam, the more times radiant heat is absorbed, and the less heat is transmitted by radiation. Similarly, the small spaces between the barbules and barbs in down feathers reduce heat flow by radiation.

So, to summarize: in down feathers, the high-volume fraction of air reduces heat loss by conduction, and the small spacing between the barbs and barbules reduces heat loss by convection and radiation. Using these ideas, we can estimate the thermal conductivity of down to be 0.036 W/mK, about 50% higher than that of air (see box 2.3).

eider down feathers:
~99% air (conductivity = 0.025 W/mK)
~1% keratin (conductivity = 0.2 W/mK)

FIGURE 2.14

Down feathers are about 99% air and 1% solid (by volume). Since air has a low thermal conductivity, down feathers do, too. [Eider down supplied by Plumeria Bay; scanning electron microscope image taken by Dr. Isaac Cabrera.]

BOX 2.3: ESTIMATE OF THERMAL CONDUCTIVITY OF DOWN FEATHERS

The total thermal conductivity of down feathers on a bird can be estimated by adding the contributions from conduction through the gas, conduction through the solid, and radiation. This calculation is somewhat approximate, but gives an idea of the amount of each contribution to the overall thermal conductivity of down.

Conduction through the gas is equal to the conductivity of air (0.025 W/mK) times the fraction of the down that is air by volume, roughly 99%:

$$0.025 \times 0.99 = 0.0248 \text{ W/mK}$$

Conduction through the solid is equal to the conductivity of solid keratin (0.2 W/mK) times the fraction of the down that is keratin by volume, roughly 1%. Since heat doesn't flow along a straight line through the keratin, but instead flows along the shaft, barbs and barbules, we multiply this value by an efficiency factor. The efficiency factor for solid conduction in foam is about 0.67; that for heat flow through the barbs and barbules of down is probably higher, because the path of heat flow through the solid is less tortuous than in a foam. For the purposes of this box, we'll use an efficiency factor of 0.8.

$$0.2 \times 0.01 \times 0.8 = 0.0016 \text{ W/mK}$$

The contribution to the total thermal conductivity from *radiation* has been measured on penguin and duck down to be between 0.007 and 0.015 W/mK (at temperatures ranging from −40°C to +10°C and for a volume fraction of solid keratin of 0.2%). At higher volume fractions of keratin, the radiative heat transfer is lower, as more radiation is absorbed by the solid. We'll use a value of 0.010 W/mK for the radiative contribution.

Adding the three contributions gives an estimate of the thermal conductivity of down:

$$0.0248 + 0.0016 + 0.010 = 0.0364 \text{ W/mK}$$

Conduction through the air within down accounts for roughly two-thirds of the estimated total thermal conductivity, conduction through the solid accounts for less than 5%, and radiation accounts for the remainder. The thermal conductivity of down from pelts of penguin chicks has been measured to be between 0.03 and 0.045 W/mK; our estimate is consistent with this.

Engineered insulations, such as foams, aerogels, and vacuum insulation panels, are designed to reduce conduction through the gas. Foam insulations are usually blown with gases that have thermal conductivities half that of air. Aerogels have very small pore sizes: on the order of 50 nanometers, less than the average distance air molecules move before hitting another air molecule, limiting conduction in the gas (1 nanometer is one millionth of a millimeter). And vacuum insulation panels, as the name suggests, have a vacuum between two panels, eliminating conduction through the gas entirely. For comparison, insulation foams have thermal conductivities of 0.02 to 0.025 W/mK, aerogels have conductivities of as low as 0.015 W/mK, and vacuum insulation panels have the lowest conductivities, around 0.005 W/mK.

The outer contour feathers of a bird also reduce heat loss. As we saw in the scanning electron micrographs of a mallard feather in chapter 1 (figure 1.2), the main shaft of a feather and the shafts of its barbs have foam-like cores that reduce heat loss along the length of the shafts.

Finally, birds often fluff themselves up into a ball in the winter (figure 2.15). When they do this, they are fluffing out their down feathers to hold more air, further reducing heat loss. And that more-spherical body shape itself reduces heat loss: spheres have the lowest surface area per volume of any shape, so in taking on that shape, birds are reducing the surface area of their bodies for heat to flow across.

Back at Cape Ann, watching the rafts of eiders, I marvel not just at their striking beauty, but also at how their life in frigid waters relies on the remarkable microscopic structures within their feathers. I marvel at the precise spacing of the barbs and barbules in their outer contour feathers that make them so water-repellent, keeping them dry. And at the flimsy-looking, microscopically branched structure of down that reduces heat loss by conduction, convection, and radiation. Their feathers are a wonder.

FIGURE 2.15

Black-capped chickadee fluffing up in the cold. [Alamy.]

REFERENCES

WATER REPELLENCY

Backcountry.com (n.d.). [Explanation of durable water-repellent fabric.] http://www.backcountry.com/explore/dwr-decoded

Bailey AI (2010). Surface and interfacial tension. *Thermopedia*. https://www.thermopedia.com/de/content/30/

Bormashenko E, Bormashenko Y, Stein T, Whyman G, and Bormashenko E (2007). Why do pigeon feathers repel water? Hydrophobicity of pennae, Cassie-Baxter wetting hypothesis and Cassie-Wenzel capillarity-induced wetting transition. *J. Colloid and Interface Science* **311**, 212–216.

Cassie ABD and Baxter S (1944). Wettability of porous surfaces. *Trans. Faraday Soc.* **40**, 546–551.

Dorr BS, Hatch JJ, and Weseloh DV (2021). Double-crested Cormorant. *Birds of the World Online*. Cornell Laboratory of Ornithology, Cornell University.

Elowson AM (1984). Spread-wing postures and the water repellency of feathers: a test of Rijke's hypothesis. *Auk* **101**, 371–383.

Gremillet D, Chauvin C, Wilson RP, Le Maho Y, and Wanless S (2005). Unusual feather structure allows partial plumage wettability in diving great cormorants *Phalacrocorax carbo*. *J. Avian Biol.* **36**, 57–63.

Rijke AM and Jesser WA (2011). The water penetration and repellency of feathers revisited. *Condor* **113**, 245–254.

THERMAL INSULATION

Chemical Rubber Company (1985). *Handbook of Chemistry and Physics*. 66th edition. Chemical Rubber Company.

Fuller ME (2015). The structure and properties of down feathers and their use in the outdoor industry. PhD thesis. School of Design, University of Leeds.

Gibson LJ and Ashby MF (1997). *Cellular Solids: Structure and Properties*. Second edition. Cambridge University Press.

Gill FB (2007). *Ornithology*. Third edition. WH Freeman.

Granta Design (n.d.). Cambridge Engineering Selector materials database.

Heinrich B (2003). *Winter World*. ECCO (Harper Collins).

Lovette IJ and Fitzpatrick JW, eds. (2016). *The Cornell Lab of Ornithology Handbook of Bird Biology*. Third edition. Wiley.

McCafferty DJ, Moncrieff JB, and Taylor IR (1997). The effect of wind speed and wetting on thermal resistance of the barn owl (*Tyto alba*) II: coat resistance. *J. Thermal Biology* **22**, 265–273.

Schuetz MA and Glicksman LR (1983). Proc. SPI 6th Int. Technical/Marketing Conference.

Stephenson R (1993). The contributions of body tissues, respiratory system and plumage to buoyancy in waterfowl. *Can. J. Zool.* **71**, 1521–1529.

Taylor JRE (1986). Thermal insulation of the down and feathers of pygoscelid penguin chicks and the unique properties of their feathers. *Auk* **103**, 160–168.

Wan X, Fan J, and Wu H (2009). Measurement of thermal radiative properties of penguin down and other fibrous materials using FTIR. *Polymer Testing* **28**, 673–679.

Williams CL, Hagelin JC, and Kooyman GL (2015). Hidden keys to survival: the type, density, pattern and functional role of emperor penguin body feathers. *Proc. Roy. Soc.* **B282**, 2015.2033.

3

BONES: LIGHT FOR FLIGHT

In all the mechanical side of anatomy nothing can be more beautiful than the construction of a vulture's metacarpal bone.
D'Arcy Thompson

When I was a graduate student at Cambridge University, working on materials with honeycomb-like or foam-like cellular structures, my thesis advisor, Mike Ashby, suggested I read D'Arcy Thompson's 1917 book *On Growth and Form*, which applies physical and mathematical principles to the growth and form of biological structures. It's a real classic, combining Thompson's extensive knowledge of zoology, mathematics, and the humanities. (The evolutionary biologist Stephen Jay Gould called it "the greatest work of prose in twentieth-century science.") The illustrations in the book are a wonder: the delicate partitions of a dragonfly's wing; the regular hexagons of a bee's honeycomb; the gorgeous geometrical cells of radiolaria, a type of small zooplankton in the ocean (just Google "radiolaria images"); and cross sections showing the interior spirals within a nautilus shell. Most fascinating to me was the drawing of a cross section of a vulture's metacarpal bone (one of the "hand" bones, today called the carpometacarpal bone) with its sandwich-like

FIGURE 3.1

A vulture's metacarpal bone (today called the carpometacarpus). [From Thompson (1961), fig. 103, based on Prochnow (1934).]

structure of outer skins of solid bone separated by a core of porous bone (figure 3.1). Engineers have long recognized that such sandwich-like structures excel at resisting bending while remaining lightweight. And with my interest in cellular materials, I was especially drawn to the structure of the porous bone, known as spongy or trabecular bone.

After I left Cambridge, I joined the Department of Civil Engineering at the University of British Columbia, where, as a freshly minted faculty member, I decided to work on two research projects, both inspired in part by the vulture's carpometacarpus. The first was on reducing the weight of engineering sandwich panels (which typically have metal- or glass-fiber- or carbon-fiber-reinforced plastic skins and engineered honeycomb, foam, or balsa cores); sandwich panels are widely used in skis, sailboat hulls, aircraft, and trucks, all applications where reducing weight is impor-tant. The second was on the mechanical behavior of trabecular bone, important in understanding the increased risk of femoral (hip) and vertebral fractures in people with osteoporosis.

And now, with the writing of this book, I've come back to bird bones again. How do the bones of birds provide the stiff, strong framework needed to support the large forces generated in flight and at the same time remain light?

BIRD SKELETONS

We can see how the skeletons of modern birds have evolved to enable flight by comparing the skeleton of *Archaeopteryx*, the oldest known bird, from 150 million years ago, with that of a pigeon (figure 3.2). *Archaeopteryx* was a crow-sized, feathered creature that shared some features with therapod dinosaurs, from which birds evolved. The name comes from the Greek *archaios*, meaning "ancient," and *pteryx*, meaning "wing." *Archaeopteryx* is thought to have been able to glide and fly—but only weakly, perhaps between trees and certainly not over long distances. Scientists have deduced this because *Archaeopteryx* hadn't yet evolved the kind of lightweight but strong and stiff skeletal structure that allows modern birds to fly so well.

One of the most obvious differences in the two skeletons is the sternum (figure 3.2, pink shading). The sternum in the pigeon is huge, while in *Archaeopteryx* it is tiny. Flight in modern birds is powered by large breast muscles that attach to the sternum—in pigeons, the breast muscles make up 20% of the weight of the bird. (Imagine a 150-pound person with 30-pound breast muscles—yikes!) The flight muscles of *Archaeopteryx* had to be tiny, corresponding to its small sternum; this is one of the reasons that paleontologists believe it to have been a weak flyer.

Several bones that are separate in *Archaeopteryx* are fused together in the pigeon (figure 3.2, brown shading). If you look at the "hand" bones of *Archaeopteryx*, for example, you'll see it has individual digits with individual bones that can move and articulate, much as the bones in our hands do—but in pigeons, some of those bones have fused together to form plate-like bones. In particular, instead of individual, cylindrical carpal and metacarpal (hand) bones for each digit, as in *Archaeopteryx*, the pigeon has a single, fused, more plate-like carpometacarpus, which allows the development of an internal sandwich structure, as we saw in the vulture carpometacarpus (figure 3.1). This sandwich structure excels at resisting bending loads imposed on the bone by the flight feathers that attach to it.

The vertebrae in the chest, where the ribs attach, are fused in pigeons. The abdominal, pelvic, and the first few tail vertebrae are fused to one another as well as to the bones of the pelvis, creating an elongated, mostly plate-like bone, known as the synsacrum, to which the abdominal, leg, and tail muscles attach. And the final few vertebrae fuse to form the pygostyle (sometimes called the "plowshare bone," because it resembles the blade of a plow), to which the tail feathers attach.

The fusion of the spine and pelvis stiffens the chest and pelvis of modern birds, so that they bend and twist less as they bear the large loads associated with the

FIGURE 3.2

The skeleton of *Archaeopteryx*, the oldest known bird, compared to that of a pigeon, showing how the skeleton has evolved in modern birds. [Image after Heilmann (1927), redrawn in Gill (2007).]

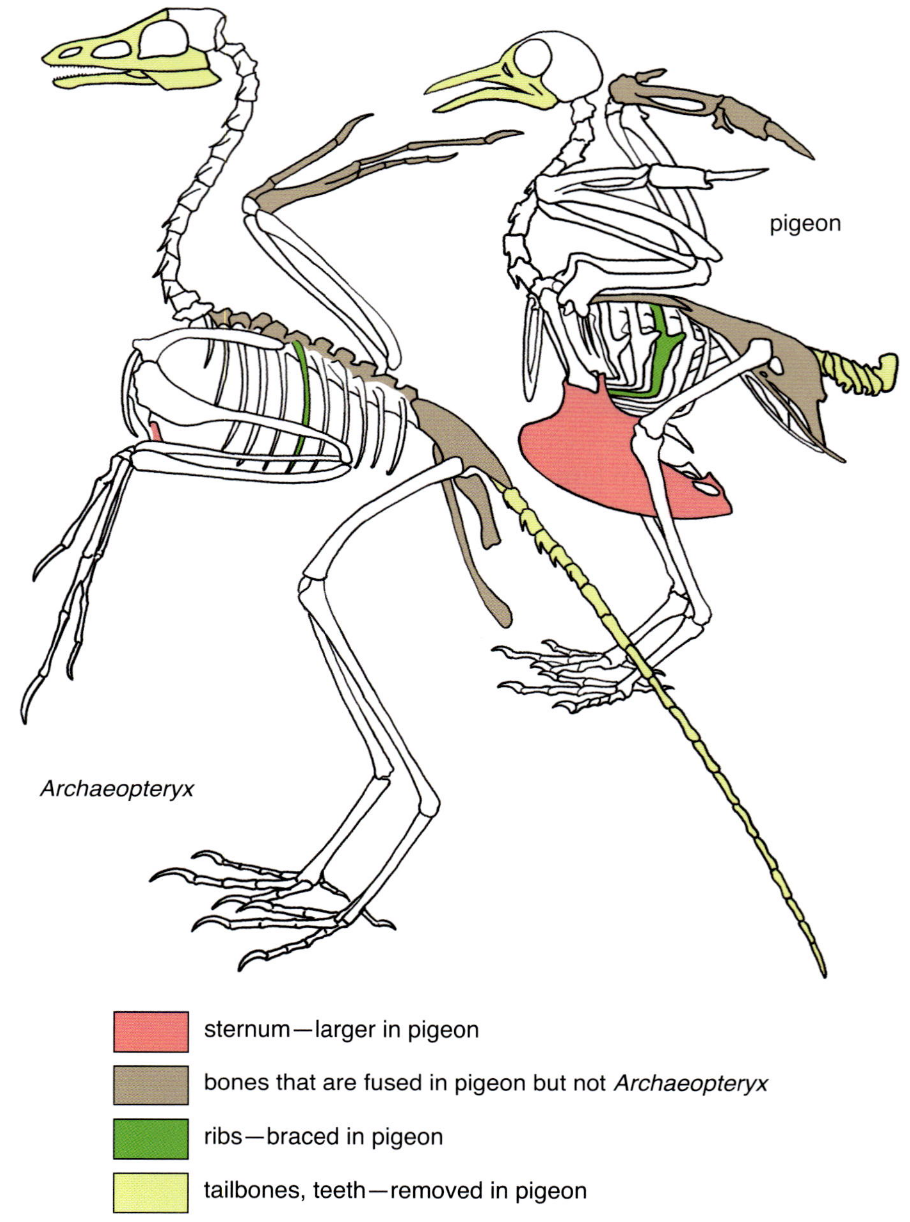

sternum—larger in pigeon

bones that are fused in pigeon but not *Archaeopteryx*

ribs—braced in pigeon

tailbones, teeth—removed in pigeon

beating of wings in flight. Without the fused bones along the spine and pelvis, birds would need large back muscles to keep their bodies sufficiently rigid, adding weight and wasting energy from tensing and relaxing them to counteract the forces generated by the flight muscles in the breast.

Most of the ribs in modern birds have two features absent in *Archaeopteryx*: a small plate that extends behind the rib at an angle of about 45° and overlaps with the adjacent rib; and a second, horizontal bone that connects the ribcage to the sternum (figure 3.2, green shading). Both features brace the ribs during flapping flight, increasing the rigidity of the ribcage.

Some bones in *Archaeopteryx* have disappeared altogether in the pigeon, reducing its weight and decreasing the amount of lift the bird needs to generate for flight (figure 3.2, yellow shading). *Archaeopteryx*'s long tail has bones running along its entire length; in the pigeon this has been replaced by a shorter stump of bone, the pygostyle, to which lightweight tail feathers attach. The heavy jawbones and teeth in *Archaeopteryx* have been replaced by a lightweight, toothless bill in the pigeon; the bill is largely made up of a core of foam-like, trabecular bone that is surrounded by a thin shell of solid bone, overlain by a sheath of keratin.

All of these features can be seen on mounted bird skeletons. In the snowy owl skeleton shown in figures 3.3 and 3.4, for example, you can see the large sternum (figure 3.3b, held up by the mounting rod), fused vertebrae, fused tail bones, fused pelvic bones, fused hand bones, as well as the small plates bracing the ribs against one another and the horizontal rib bones connecting to the sternum. The skeleton also reveals a hidden feature of owls: although they often look like they have barely any neck, they actually have quite long necks relative to the size of their bodies (figure 3.3). Nearly all mammals, from mice to humans to giraffes, have 7 neck vertebrae—but owls have 14, which is, in part, why they can rotate their heads through three quarters of a circle, or 270°. Some birds have even more vertebrae in their necks. Swans have 25.

(a)

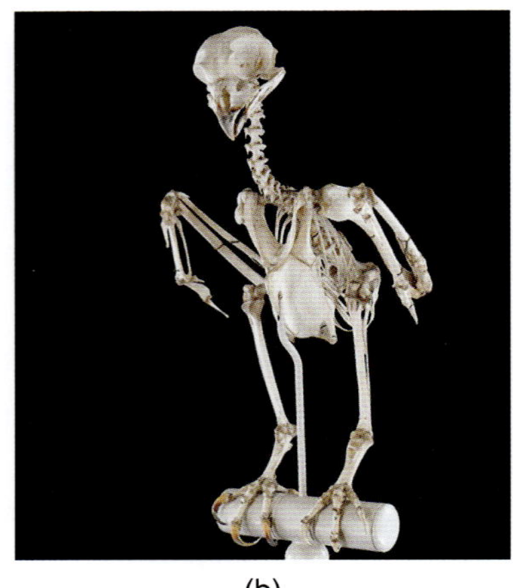

(b)

FIGURE 3.3

(a) Snowy owl and (b) snowy owl skeleton. [(a) Alamy; (b) Museum of Comparative Zoology, Harvard University, specimen 341413, © President and Fellows of Harvard College.]

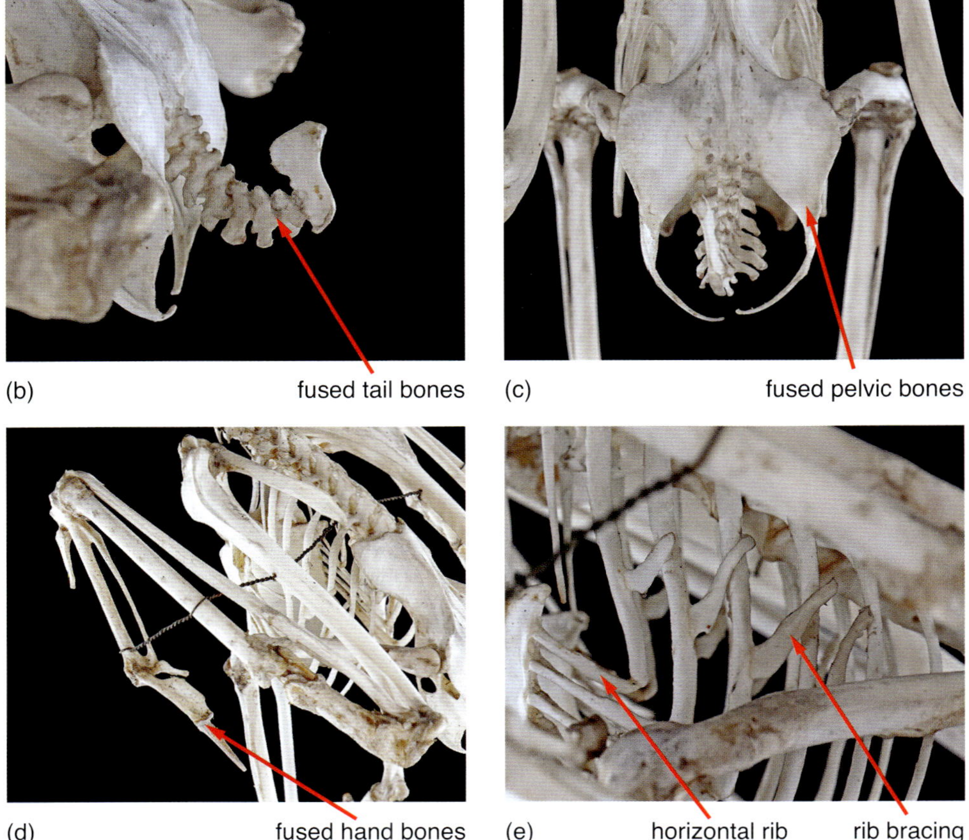

FIGURE 3.4

Snowy owl skeleton details: (a) fused thoracic vertebrae, (b) fused tail bones, (c) fused pelvic bones, (d) fused hand bones, (e) rib bracing. [Museum of Comparative Zoology, Harvard University, specimen 341413, © President and Fellows of Harvard College.]

As we saw in chapters 1 and 2, the internal structure of feathers gives rise to their remarkable properties: the iridescence of hummingbird feathers, the water repellency of duck feathers, the thermal insulation of eider down. We'll now look at the internal structures of bird bones to see how they produce the remarkable properties that allow them to resist the large loads imposed on them by flight while remaining lightweight.

SANDWICH STRUCTURES

We've already seen the sandwich structure of the vulture carpometacarpus bone, with its solid top and bottom plates of bone separated by trabecular bone (figure 3.1). Sandwich structures are also on display in the sternum and the synsacrum (figures 3.5 and 3.6). The sternum has a large "keel" perpendicular to its base, to which the pigeon's huge flight muscles attach (figure 3.5). The fusion of the pelvic and lower vertebral bones in the synsacrum is clearly visible in figure 3.6. The entire sandwich of both bones is thin (about 1 mm thick), with outer solid skins of about 0.1 mm or 100 microns (μm). (As a point of comparison, human hair is about 50 microns in diameter.) That's remarkably thin: living bone cells, which produce the solid bony material, are about 30 microns across, so the outer skins of these sandwich structures are only just slightly larger than bone cells themselves. Also amazing is the precise interior structure of the bone, with its rods organized nearly perpendicular to the outer skins of the sandwich. When I look at these bones, they seem almost like works of sculpture—they're really beautiful.

The sandwich structure of these two bones, along with the metacarpal bone, makes them highly resistant to forces that might bend them. These include large aerodynamic forces that are transmitted to the metacarpal bone via the flight feathers attached to it, the constantly shifting loads of flight on each wing for the sternum, and the alternating left-right loads of walking for the synsacrum. And sandwich structures excel at resisting bending while also weighing less than solid bone for any given stiffness (how much load it takes to produce a given deflection) or strength (how much load it takes to break the bone).

Let's get into a few technical details. First, let's look at a rubber beam, with gridlines marked on its side (figure 3.7). When I bend the beam, the vertical lines rotate, becoming closer together at the top and further apart at the bottom. The top of the beam is squeezed together in compression, and the bottom is stretched

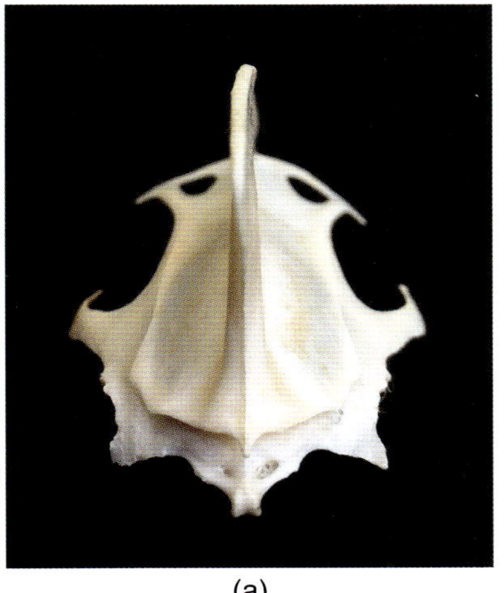

(a)

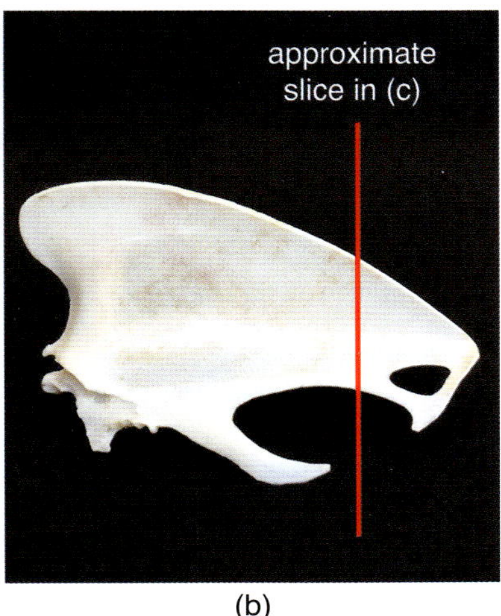

(b)

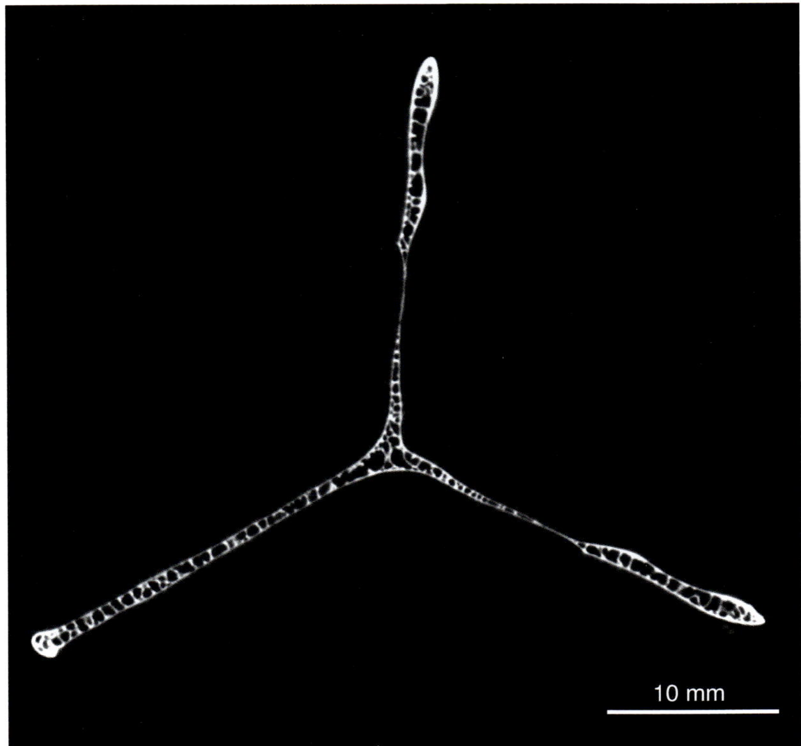

(c)

FIGURE 3.5

(a, b) Pigeon sternum and (c) section of a micro-computed tomography scan showing the sandwich structure of the sternum, with solid outer skins separated by tiny ribs. [Concord Field Station, specimen 3887; micro-CT image by L Fahn-Lai, slice 950; Department of Organismal and Evolutionary Biology, Harvard University.]

apart in tension. The horizontal plane of material at the very center of the beam (the horizontal line labeled "neutral axis") neither compresses nor stretches; it sees no force along the length of the beam during bending. The compressive force in the top half of the beam increases from zero at the center to the maximum value at the top of the beam. Likewise, the tensile force in the bottom half of the beam increases from zero at the center to the maximum value at the bottom of the beam. Since the internal tensile and compressive forces in a beam are highest at the top and bottom, it makes sense that more material is needed there, to resist these larger forces.

But these are not the only forces within the beam. In figure 3.8, we have a beam made up of several thin layers that are not bonded together. When I bend the beam, the layers slide over one another. If I bonded the layers together and then bent the beam, this sliding of one layer over another would be resisted, developing shear forces within the layers, acting parallel to the layers. In a solid beam, too, there are shear forces within it. Typically, the shear forces within a beam are much, much smaller than the tensile and compressive forces at the top and bottom of the beam.

To summarize, without going into all the mathematical details: in a sandwich beam, nearly all the tensile and compressive forces along the length of the beam are carried by the solid top and bottom faces, while nearly all the much smaller shear forces are carried by the lightweight, porous core. Steel I-beams that you might see in a steel framework of a high-rise building on a construction site work in a similar way: the outer flanges carry the tensile and compressive loads while the narrow web separating them carries the shear forces (figure 3.9).

If we simplify the geometry for a generic bird bone sandwich, we can get an idea of just how effective its structure is in reducing the weight of a beam. To do that, let's straighten out the sandwich beam of bone (with a rectangular cross section), assume that the inner core of trabecular bone is foam-like (as it is in some bird sandwich structures), and then compare its weight to that of a solid bone beam of the same length, width, and stiffness—that is, let's assume that the sandwich and solid bone beams deflect the same amount under the same load (figure 3.10).

The comparison yields remarkable results. With the help of micro-computed tomography scans (figure 3.5), we can estimate that a bird bone sandwich has solid skins that are roughly 0.1 mm thick. We'll assume that the trabecular bone core is roughly 0.8 mm thick and only 10% dense (that is, 90% of the volume is air) (figure 3.11). For the same stiffness, the solid bone beam needs a depth of 0.79 mm, which makes it dramatically heavier. The bone sandwich beam weighs only 38% as much as the solid bone beam, a remarkable savings in weight.

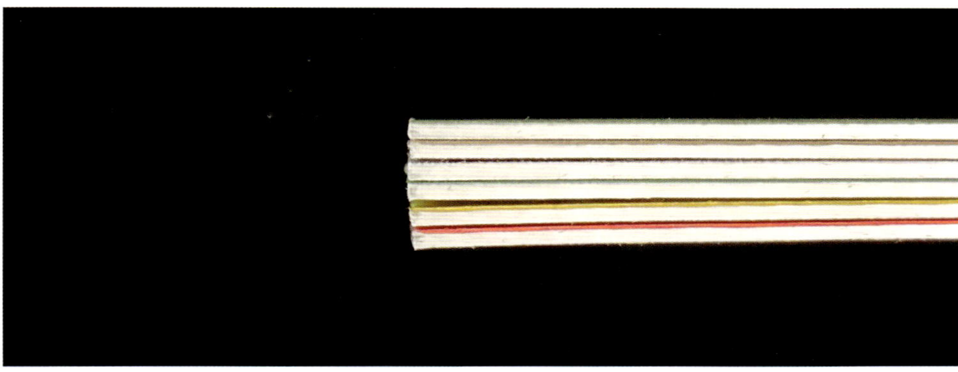

FIGURE 3.8

(a) A beam made up of several sheets of card, unloaded. (b) When bent, the sheets of card slide over one another, in shear.

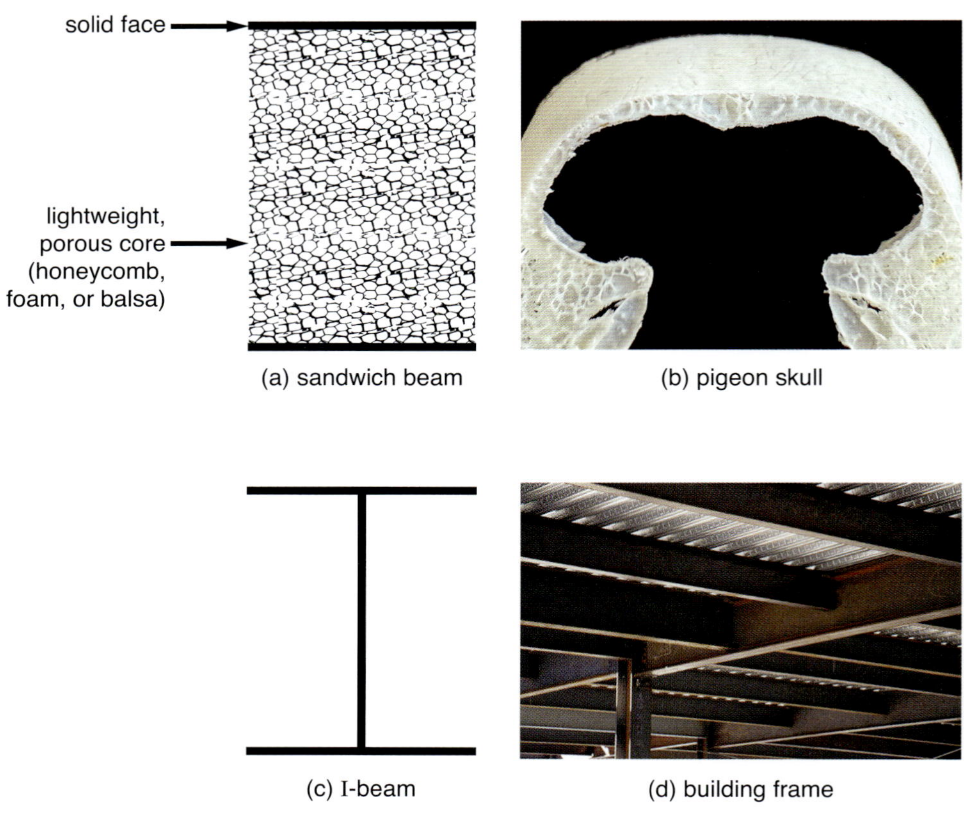

FIGURE 3.9

(a) Schematic of a sandwich beam, (b) pigeon skull with a sandwich structure, (c) schematic of an I-beam, and (d) steel I-beams in a structural framework. [(b) Concord Field Station, Harvard University, specimen CFS 3907; (d) Alamy.]

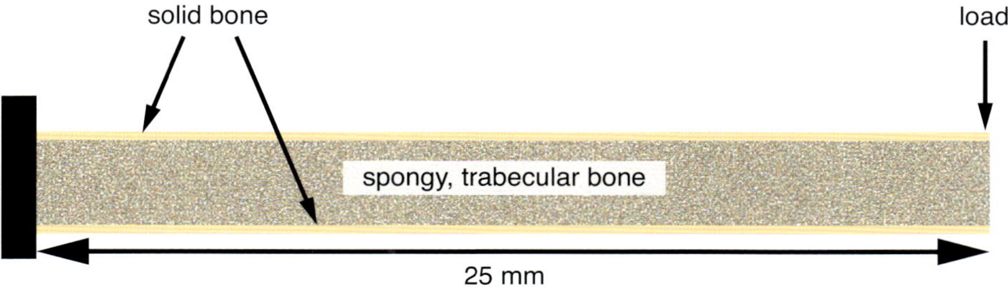

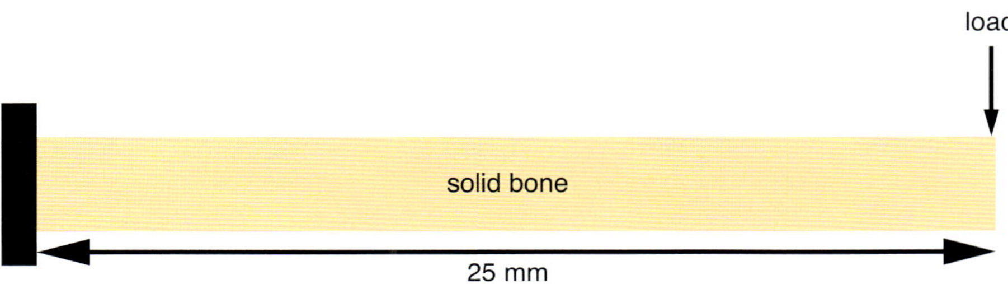

same length, width,
same **stiffness** (or load/deflection)

weight of each beam = ?

FIGURE 3.10

Comparison of a bone sandwich beam with a hypothetical solid bone beam. The two beams have the same length, width, and stiffness (they deflect the same amount under a given load).

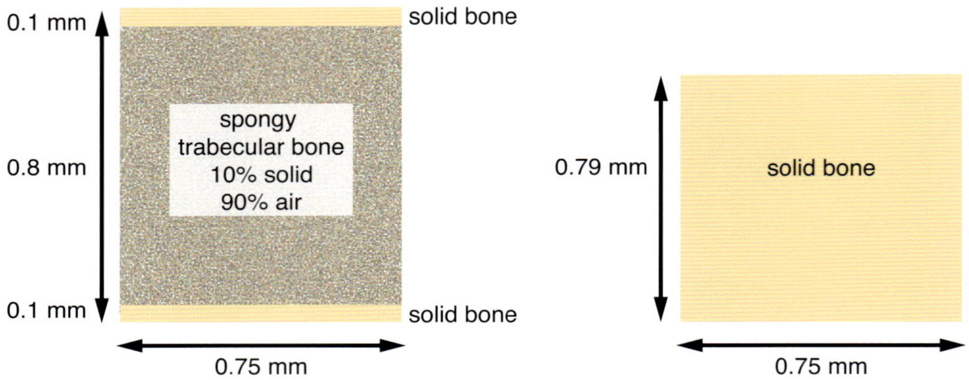

FIGURE 3.11

The cross sections of the bone sandwich beam and the solid bone beam, for the same stiffness. The bone sandwich beam is 38% of the weight of the solid bone beam.

Similarly, when we calculate the weights of sandwich and solid beams of bird bone that fail at the same load (with all else equal, as above), we discover that the sandwich beam weighs only 43% as much as the solid beam (figures 3.12 and 3.13). I haven't tried to optimize anything in these calculations, but even so, as you can see, sandwich structures in bird bones are impressively good at reducing weight while providing stiffness and strength.

Sandwich structures appear widely in nature—in the leaves of irises, cattails, and grasses, for example. Each of these plants is a monocotyledon, which is a fancy way of saying that a seed gives rise to a single stalkless leaf (figure 3.14). How does a single insubstantial leaf of any one of these plants manage to grow straight up and then remain standing almost vertically? You've probably never thought about this before, but by now you've already guessed the answer: sandwich structures. Iris leaves have nearly solid fibers that run longitudinally along their top and bottom surfaces, and that act a little like the glass fibers in a glass-fiber-reinforced composite, separated by a core of foam-like tissue. Cattail leaves, too, have bundles of nearly solid, longitudinal fibers on their top and bottom surfaces, separated by a series of parallel ribs, each one about a millimeter high, similar to the pigeon sternum and synsacrum sandwich structures (figures 3.5, 3.6).

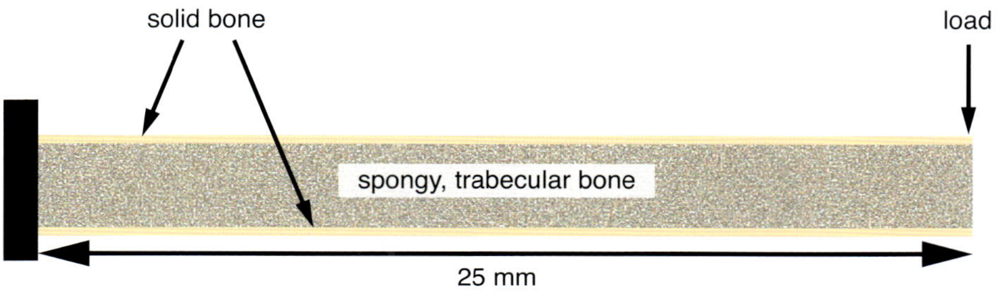

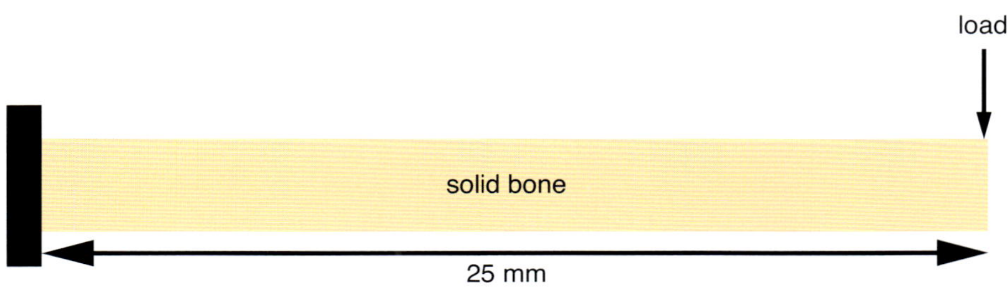

same length, width, same **failure load**

weight of each beam = ?

FIGURE 3.12

Comparison of a bone sandwich beam with a hypothetical solid bone beam. The two beams have the same length, width, and strength (failure load).

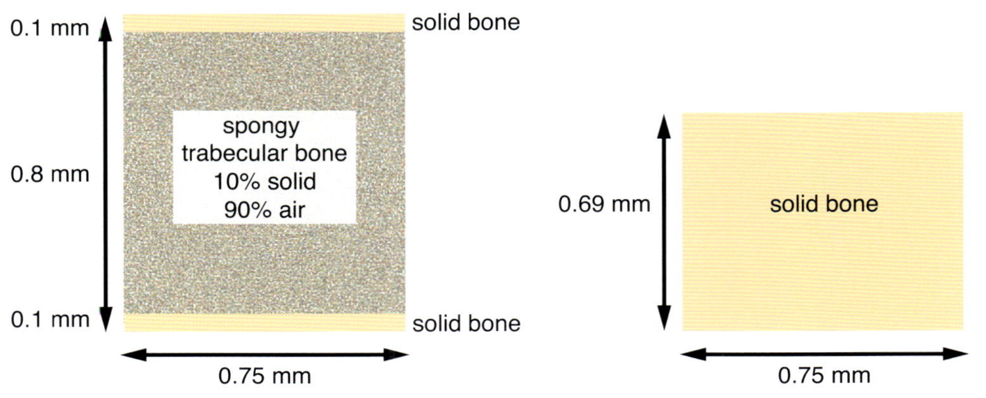

for same length, width, **failure load**

sandwich beam is **43% of the weight** of the solid beam

FIGURE 3.13

The cross sections of the bone sandwich beam and the solid bone beam, for the same failure load. The bone sandwich beam is 43% of the weight of the solid bone beam.

FIGURE 3.14

Sandwich structures in nature. Scanning electron microscope images of (a) an iris leaf, with its solid fibers at the top and bottom of the leaf and foam-like core in the middle of the leaf, and (b) a cattail leaf, with the top and bottom skins separated by ribs. [Photos by Don Galler and Lorna Gibson.]

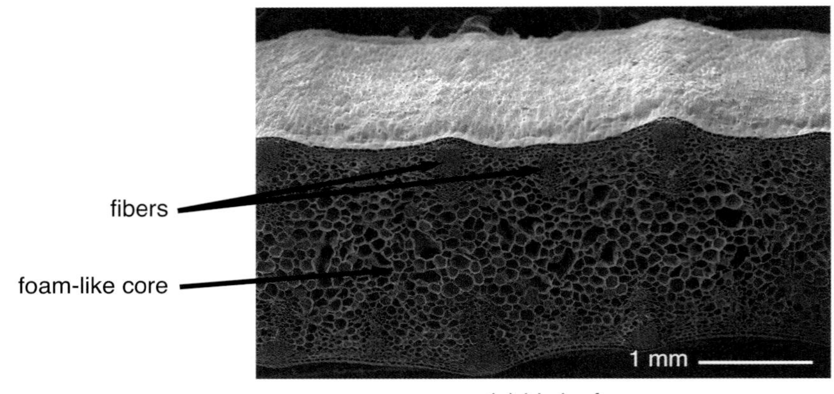

(a) iris leaf

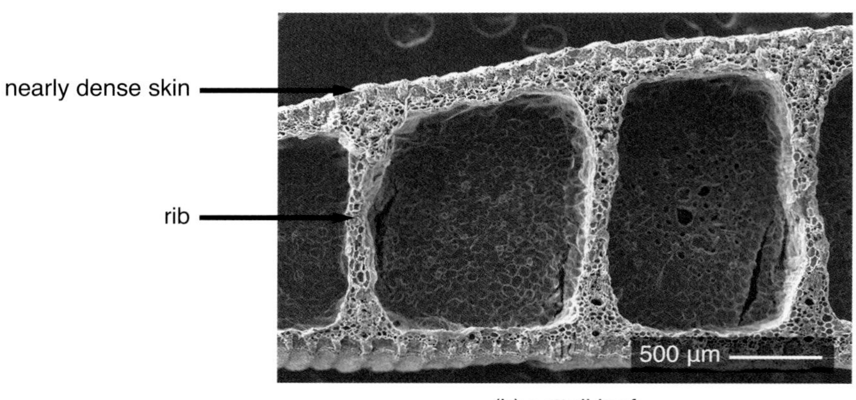

(b) cattail leaf

In the mid-twentieth century, as engineers grappled with the demands of designing airplanes, they came to appreciate the value of sandwich structures. They first used them during World War II in the Mosquito airplane, the fuselage of which had a sandwich structure that consisted of birch skins and a balsa-wood core (figure 3.15). Thanks in part to the low weight of its sandwich fuselage, the Mosquito was able to fly at speeds of up to 400 mph (650 km/hr), which made it the fastest aircraft in service for most of the war. And sandwich panels are still widely used in aircraft components, although today they typically have carbon fiber skins and engineered honeycomb cores; almost half of the outer skin of Boeing 757s and 767s are made of sandwich panels. Sandwich panels have long been used in helicopter rotor blades, too. They're also used in wind turbine blades, which today can be up to 260 feet (80 m) long, as they're light enough to accelerate easily in the wind.

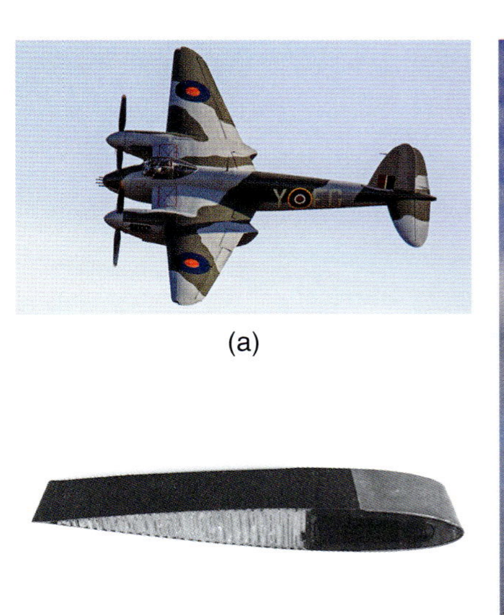
(a)

(b)

(c)

FIGURE 3.15

Sandwich structures in engineering: (a) World War II Mosquito airplane, (b) cross section of a helicopter rotor blade, (c) wind turbine rotor blades. [(a, c) Alamy; (b) Gibson and Ashby (1997), © Cambridge University Press, Reprinted with permission.]

CYLINDRICAL LONG BONES

I'm meeting with Vanya Rohwer, the curator of birds and mammals at Cornell University's Museum of Vertebrates, eager to see their bird collection. Vanya is full of enthusiasm and delightedly shows me some highlights: the skull of a hornbill, with its huge, curved, pale yellow bill and the casque on top of its head, with the interior trabecular bone showing in an area where the outer layer of bone has broken away; a giant synsacrum from a common ostrich, with its fused vertebrae and pelvic bones; boxes of bones from individual skeletons. Then he hands me the long humerus (upper wing bone) from an albatross. As I grasp it, I laugh in delight. Even though I'm an engineer and know that bird bones are light, I'm still surprised at how stunningly light it is.

The long bones of the wings and legs of birds, like the long bones of mammals, are roughly cylindrical tubes with enlarged ends. But in birds the upper long bones of the wing (the humeri) are typically filled with air rather than marrow. This reduces weight, of course. But, as Rohwer explained, to my astonishment, the hollows within the humeri of birds—in addition to several other bones, including the neck vertebrae and sternum—also serve as part of their respiratory system: they connect with air sacs in the chest and groin areas to supplement the reservoir of air that those sacs supply to the lungs. In their hollowness, these bones are an exception; most bird bones are filled with marrow, which is necessary for the production of blood cells, that then flow throughout the bird's body.

In the long bones of birds, the wall of the cylindrical section tends to be very thin: thinner, relative to the diameter, than in mammals. Figure 3.16 shows the upper wing bone (the humerus) of a tundra swan, and one of the foot bones (the metatarsal) of a white-tailed deer. Both bones are of similar length and diameter, but the thickness of the swan humerus is just 4% of its diameter, while that of the deer metatarsal is 26% of its diameter.

In general, just how thin are the long bones in birds? In raptors, owls, ravens, albatrosses, swans, and pelicans, the thicknesses of the humerus and femur have been measured to be 5–15% of their diameters (table 3.1). Measurements on fossil ulna and femur of *Archaeopteryx* have values of thickness to diameter of 13–20%, a little higher than typical in modern birds. The corresponding values for land mammals (taken from a somewhat random selection of species) ranges from 14% to 32%. Roughly speaking, this means that the long bones in birds are about one-half to one-third as thick as those in land mammals, relative to their outer diameter.

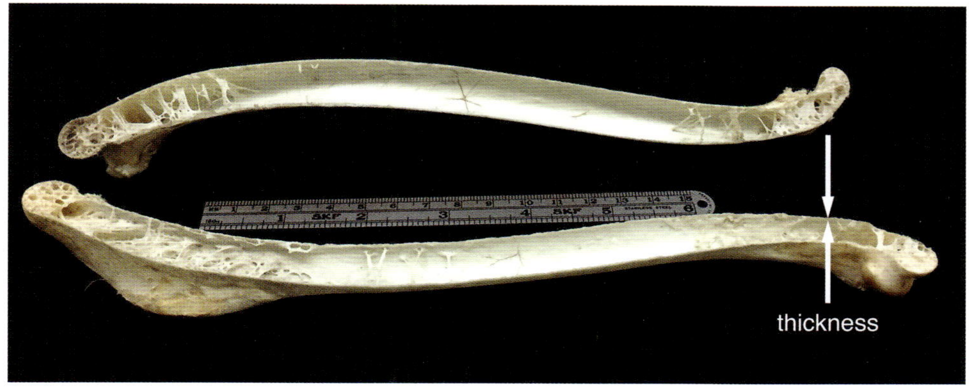

(a) tundra swan humerus:
length = 250 mm, wall thickness = 0.6 mm, diameter = 15 mm, t/D = 0.04

(b) white-tailed deer metatarsal:
length = 263 mm, wall thickness = 5.1 mm, diameter = 19.8 mm, t/D = 0.26

FIGURE 3.16

Long bones in birds have a much lower ratio of wall thickness to diameter than mammal bones. The (a) tundra swan humerus and (b) white-tailed deer metatarsal are nearly the same length and diameter, but the swan bone is much thinner-walled than the deer bone. [Museum of Comparative Zoology, Harvard University, Ornithology and Mammalogy Departments, specimens 364154 and BOM3542, © President and Fellows of Harvard College.]

Table 3.1: Thickness of bones of different species, relative to their diameter

Species	Thickness Relative to Diameter
Most birds	5–15%
Archaeopteryx	13–20%
Land mammals	14–32%
Diving birds	14–32%

Data from Habib and Ruff (2008); Simons et al. (2011); Bühler (1992); Currey and Alexander (1985).

That said, there are some exceptions. The leg bones of flightless land-based birds, such as the ostrich and the even larger (but now extinct) moa, have thickness-to-diameter ratios in the range of land mammals (for instance, the thickness-to-diameter ratio of an ostrich femur is 14% while that of a moa femur is 27%). The long bones of diving birds, which need to counteract the buoyancy of their feathers, also have thickness-to-diameter ratios that are similar to those of land mammals (and none are air-filled). Interestingly, in divers that "fly" through the water with their wings, such as alcids (e.g., murres, puffins) and penguins, the humerus is thicker-walled than the femur (relative to the diameter), as might be expected from the higher loads on their wings.

How much do the relatively thin-walled bones of a bird reduce weight? If we compare two cylinders of the same length and stiffness (same deflection for the same load), one with a thickness-to-diameter ratio of 10% (corresponding to bird bones) and the other with a thickness-to-diameter ratio of 20% (corresponding to mammals), we find that the thinner-walled bone weighs 68% as much as the thicker-walled bone (figure 3.17). The weight savings is the same for cylinders of equal stiffness in bending or torsion (twisting) and is similar for cylinders of equal strength in bending (figure 3.18) or torsion. In general, the thinner the wall, relative to its diameter, the lower the weight.

So why not an even thinner tube wall? Because as the tube is bent, one side is compressed and the other is stretched (figure 3.7)—and very thin-walled tubes can fail by kinking or wrinkling on the side of the tube in compression. That's what happens when you bend a drinking straw: if you bend it enough, the straw kinks on the compressive side.

Theoretically, for a perfect cylinder with uniform thickness and constant bone properties, what's the lowest value of thickness/diameter (t/D) to avoid kinking in hollow long bones (before reaching the compressive strength of the bone itself)?

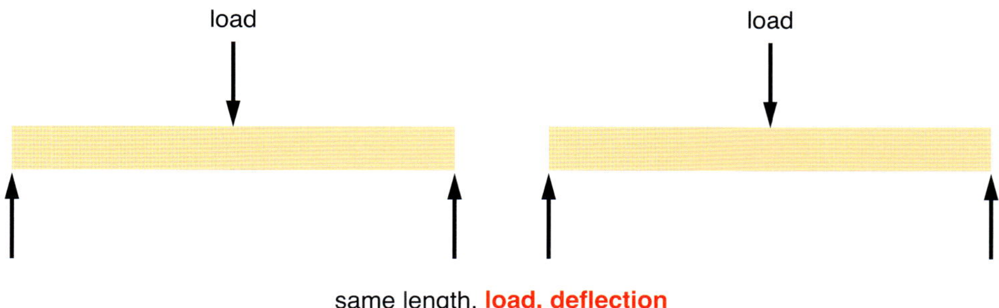

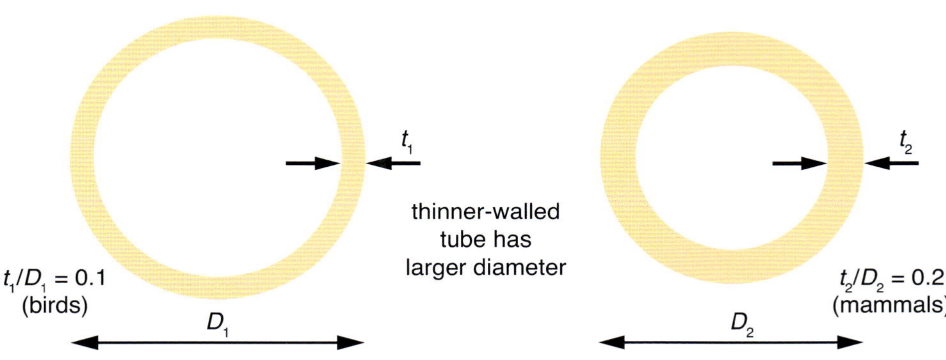

bird-like tube is **68% of the weight** of mammal-like tube

FIGURE 3.17

Comparison of hollow tubes with thickness/diameter ratios typical of birds ($t_1/D_1 = 0.1$) and mammals ($t_2/D_2 = 0.2$), with the same length and stiffness (they deflect the same amount under a given load). The one typical of birds is 68% of the weight of that typical of mammals.

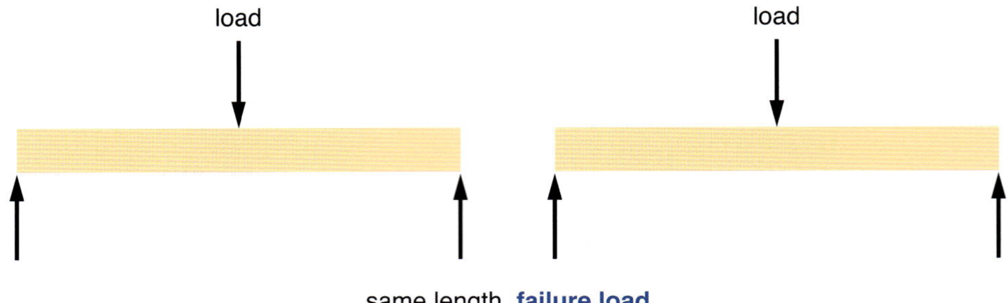

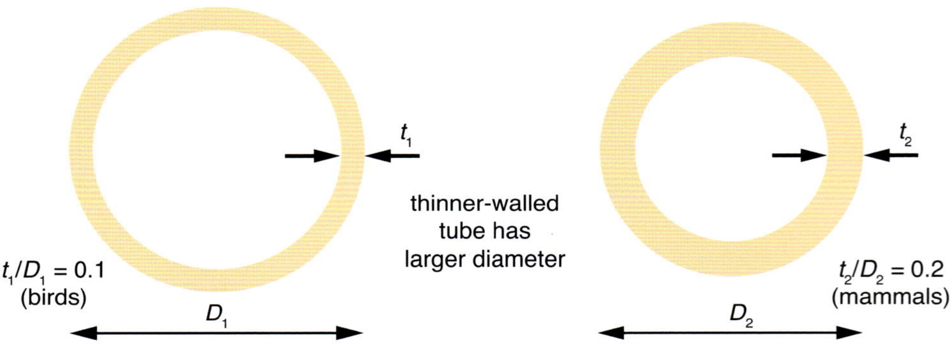

bird-like tube is **73% of the weight** of mammal-like tube

FIGURE 3.18

Comparison of hollow tubes with thickness/diameter ratios typical of birds ($t_1/D_1 = 0.1$) and mammals ($t_2/D_2 = 0.2$), with the same length and failure load. The one typical of birds is 73% of the weight of that typical of mammals.

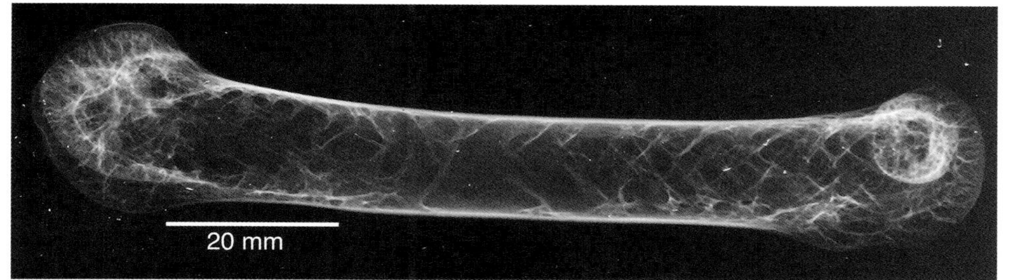

FIGURE 3.19

Radiograph of a femur of a northern ground hornbill, showing the interior bony lattice that helps prevent kinking of the tube when bent. [American Museum of Natural History, specimen AMNH 1770, with permission.]

About 1.5%. In practice, variations in the wall thickness or in the bone properties mean that the lowest value of t/D to avoid kinking is actually a little higher, about 3%. The value of t/D for the humerus of the tundra swan, 4% (figure 3.16), is just above this lower limit for a hollow long bone.

One way to increase the load at which kinking occurs is to provide additional interior support to the tube, which acts as a spring pushing back against kinking. The long bones of some birds have an interior bony lattice that helps resist kinking. Figure 3.19 shows a radiograph of the femur of a northern ground hornbill (t/D = 0.04), with a rather lovely interior latticework of bone supporting the outer tube.

Looking out my study window, I see a line of six mute swans flying low over the neighborhood houses, heading for Jamaica Pond. They are huge, one of the largest waterfowl in North America, with a 7-foot wingspan. In flight they are impressive: necks straight out, pure white, wings beating steadily. Swans are strong fliers, capable of reaching speeds up to 50 mph (80 km/hr). Critical to their ability to fly are their stiff, strong, yet lightweight bones.

As I watch them fly by, I think of how the material in their bones is arranged so perfectly, just where it is most needed. In the sternum, synsacrum, and manus (or "hand" bone), the sandwich structure of the bones excels at resisting bending, with solid bone at the top and bottom of the sandwich resisting compressive and tensile loads while the network of internal struts carries the shear loads. And the very thin walls of the long bones such as the humerus and femur allow them to resist bending and twisting at a much lower weight than if they were thicker-walled. Every time I see one flying, I stop in awe and watch.

REFERENCES

Bühler P (1972). Sandwich structure in the skull capsules of various birds—The principle of lightweight structures in organisms. In *Information of the Institute for Lightweight Structures*, vol. IL4, p41–50. University of Stuttgart.

Bühler P (1992). Light bones in birds. *Los Angeles Mus. Nat. Hist. Sci. Ser.* **36**, 385–394.

Currey JD (2002). *Bones: Structure and Mechanics*. Princeton University Press.

Currey JD and Alexander RM (1985). The thickness of the walls of tubular bones. *J. Zool. Lond.* **206**, 453–468.

de Kok-Mercado F, Habib M, Phelps T, Gregg L, and Gailloud P (2013). Adaptations of the owl's cervical & cephalic arteries in relation to extreme neck rotation. *Science* 339, 514–515. DOI:10.1126/science.339.6119.514

Gibson LJ, Ashby MF, and Harley BA (2010). *Cellular Materials in Nature and Medicine*. Cambridge University Press.

Gill FB (2007). *Ornithology*. Third edition. WH Freeman.

Habib MB and Ruff CB (2008). The effects of locomotion on the structural characteristics of avian limb bones. *Zoological Journal of the Linnean Society* **153**, 601–624.

Heilmann G (1927). *The Origin of Birds*. D Appleton.

Leahy C (2004). *The Birdwatcher's Companion to North American Birdlife*. Princeton University Press.

Prochnow O (1934). *Formenkunst der Natur* Ernst Wasmuth Vlg.

Proctor NS and Lynch PJ (1993). *Manual of Ornithology: Avian Structure and Function*. Yale University Press.

Rogers RR and LaBarbera M (1993). Contribution of internal bony trabeculae to the mechanical properties of the humerus of the pigeon (*Columba livia*). *J. Zool. Lond.* **230**, 433–441.

Simons ELR, Hieronymus TL, and O'Connor PM (2011). Cross sectional geometry of the forelimb skeleton and flight mode in Pelecaniform birds. *J. Morphology* **272**, 958–971.

Thompson DW (1961). *On Growth and Form*. Abridged edition, edited by Bonner JT, foreword by Gould SJ. Cambridge University Press.

Vestas Wind Systems 3MW Platform Brochure. https://www.northlandpower.com/en/resources-General/ProjectDocuments/BallHill/Appendix%20A%20-%20Vestas%20126,%203.45%20MW%20Series%20Wind%20Turbine%20Brochure.pdf

von Grouw K (2013). *The Unfeathered Bird*. Princeton University Press.

Wennberg D (2011). Light-weighting methodology in rail vehicle design through introduction of load carrying sandwich panels. Licentiate thesis, Department of Aeronautical and Vehicle Engineering, KTH (Royal Institute of Technology), Stockholm, Sweden.

Winchester S (2012). *Skulls: An Exploration of Alan Dudley's Curious Collection*. Black Dog and Leventhal Publishers.

4

BILLS

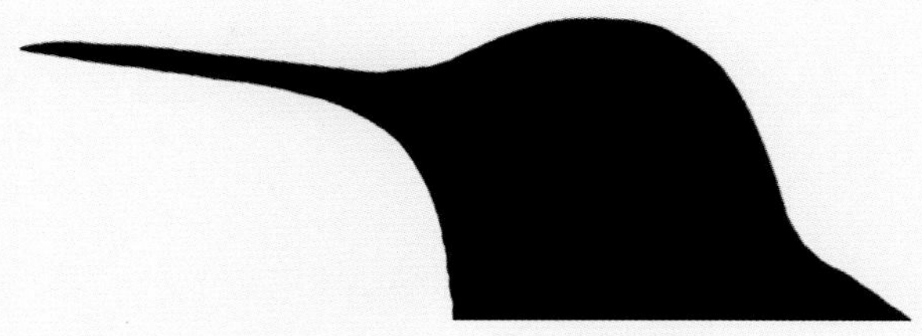

Beaks are to birds what hands are to us.
Jonathan Weiner

I'm at the Wild Center, a nature sanctuary and natural history museum in the Adirondacks, looking at a twice-life-size wooden model of a pileated woodpecker mounted on a tree trunk, its large bill prominently on display (figure 4.1). Across from it, up against the trunk, is a giant Swiss Army knife opened up to reveal its long and short blades and its bottle and can openers, all jutting into the tree. And suddenly I get it: bills are the Swiss Army knives, the multipurpose tools, of the bird world. Having given up their "hands" to form part of their wings for flight, birds have to do a lot of their handiwork with their bills, or beaks (the terms are used interchangeably).

Feeding is the main function of bills, and the one we'll focus on in this chapter, but it's not the only one. Most birds use their bills to build nests, from the common nests of sparrows to the huge, condo-style, hanging nests of weaver birds that accommodate multiple families. Birds use their bills to preen their feathers, to maintain the integrity of the vane and to spread wax from the preen gland (at the base of the tail) over their feathers to prevent them from drying out and becoming brittle.

FIGURE 4.1

Bills are the Swiss Army knives of the bird world. Display from the Wild Center, Tupper Lake, NY.

They use their bills for defense against predators. Hummingbirds, in spite of their small size, are famously aggressive, using their bills to attack and drive off rivals. Rival male northern flickers use their bills to duel in front of a prospective mate in the spring. Woodpeckers use their bills to make sound, drumming on hollow branches (or, annoyingly, on the drainpipe of a house) to declare their territory and readiness to mate. Somewhat surprisingly, toucans use their spectacularly large bills to help regulate their temperature, by controlling the flow of blood through a network of vessels close to the surface of their bills. Most remarkably, New Caledonian crows use their bills to make and manipulate tools to obtain food.

Bills come in a wide array of sizes and shapes (figure 4.2), from hummingbirds' long bills (for probing into the base of a flower to collect the nectar within); to the larger, robust bills of finches (for breaking open seeds); to the short, needle-like bill of swallows (for picking insects out of the air); to the sharply hooked bills of red-tailed hawks and other raptors (for tearing into meat); to the chisel-like bills of woodpeckers (for drilling holes into trees to capture insects or extract sap).

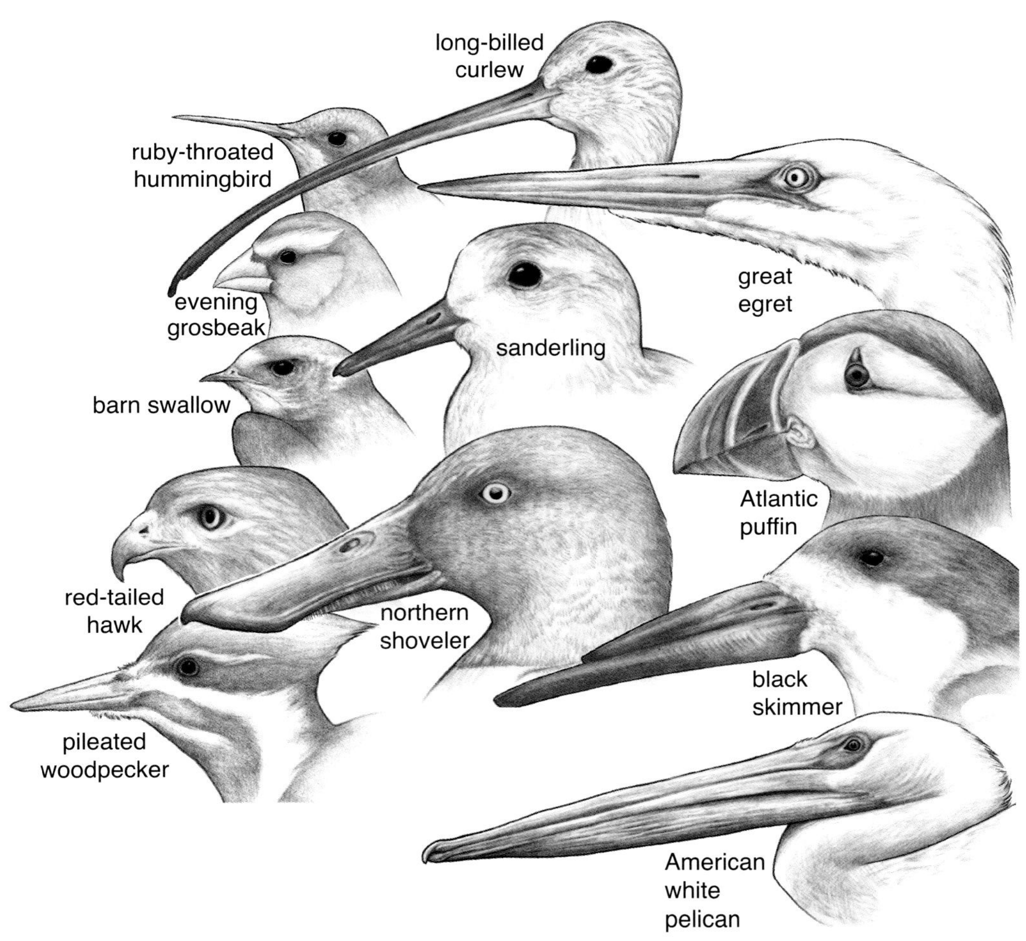

FIGURE 4.2

Bird bills have a variety of shapes to facilitate eating different foods. [After Proctor and Lynch (1993).]

Birds that feed on invertebrates in the water and along the shore have even more varied bill shapes. Shorebirds use their bills to probe into sand and pluck out insects and invertebrates; their bills vary from the long, curved bills of the long-billed curlew and whimbrel to the relatively short, straight bills of sanderlings and red knots. Dabbling ducks, such as mallards, use their broad bills to gather invertebrates and plants along the shoreline and at shallow depths. And filter feeders, like northern shovelers and flamingos, have unusually shaped bills and tongues with hair-like projections around the edges that allow water to flow through while retaining microorganisms within the bill.

Fish-eaters use different strategies and have bills to match. Waders, such as herons and egrets, have long, dagger-like bills to spear fish. Some diving birds, like mergansers and cormorants, have long, thin, serrated, hooked bills that they use to grab and hold onto fish. Others, like the alcids, have broader bills; puffins can hold several small fish in their bills at the same time. And some birds fly low over the surface of water to scoop or snatch fish: skimmers skim the water with the longer, lower part of their bill, which snaps shut when it touches a fish, and pelicans do so with their pouch-like bills.

That's a lot of different sizes and shapes of bills, too many to discuss in detail. Instead, we'll look at a few examples of how bills work. How do woodpeckers peck without suffering brain injury? How do hummingbirds draw liquid nectar into their mouths? How do shorebirds find prey beneath the surface of the sand on a beach? How do phalaropes—another shorebird—draw beads of water containing plankton up into their mouths?

WOODPECKERS: BILLS FOR PECKING

It's mid-October, and I've come to the Massachusetts Audubon Broadmoor sanctuary, in Natick, Massachusetts, to go for a walk and do a little leaf-peeping. The trees are spectacular, vermillion and gold, especially along the banks of the Charles River, which weaves its way some 20 miles from here before it gets to Boston and empties out into the Atlantic. Walking along, I hear a slow, loud, determined pecking quite close to me—so loud that I think that perhaps it's a pileated woodpecker and stop to look closely among the trees. And there it is: hopping up a tree trunk, pecking different spots as it goes, then flying across the path in front of me, landing on another tree trunk. After a couple of minutes, a second pileated woodpecker flies in, landing on a trunk near the first. I watch them for about ten minutes, flying with their undulating pattern between trees, landing on trunks, pecking as they work their way along the trunk. At one point they're on the same tree trunk, facing each other. It's always exciting just to see one, so a pair is a real treat!

Woodpeckers eat insects from beneath the bark of decaying trees, and to do that they need a sturdy, chisel-like bill. They can actually hear insects moving around beneath the bark, which is how they know where to peck. Woodpeckers have two remarkable anatomical arrangements for capturing insects within the holes that they make. The first is the bony hyoid, connected to the tongue, that wraps around the back of the skull and anchors at the front, often at the base of a nostril. Using neighboring muscles, woodpeckers can contract the hyoid, which then enables them to scoot their tongue far out and into the pecked hole (figure 4.3). The second arrangement involves rear-facing bristles at the tip of the tongue, which enable the woodpecker to hold onto insects and larvae as the tongue is drawn back into the mouth.

The acorn woodpecker uses its chisel-like bill not to peck holes to capture insects but instead to drill cavities for storing acorns (figure 4.4). I think of acorn woodpeckers as the champions of pecking: in a social group of a dozen or so, they can store thousands of acorns in decaying tree trunks. These storage areas are known as granaries.

Woodpeckers also use their bills to carve out cavity nests for themselves, usually in decaying tree trunks. The cavity nests of large woodpeckers are impressive: pileated woodpecker nests can be up to two feet deep and take three to six weeks to excavate. The ivory-billed woodpecker nest in a section of a bald cypress tree trunk in the collection of the Museum of Comparative Zoology at Harvard University is about 8 inches (200 mm) in diameter and 21 inches (530 mm) deep. That requires a lot of pecking.

But woodpeckers don't only excavate trees. In June of 1995, famously, a pair of yellow-shafted northern flickers, possibly scouting out a nest site, pecked several holes in the foam surrounding the fuel tank of the space shuttle *Discovery* as it sat on the launch pad. The foam on the tanks insulates them to keep the liquid hydrogen fuel inside at −423°F (−253°C). Needless to say, the holes delayed the launch. NASA responded by convening the Bird Investigation Review and Deterrent (BIRD) committee, which recommended removing palms and dead trees from the area around the launch pad as well as some scare tactics: plastic owls, water sprays, and "scary eye" balloons near the tank.

FIGURE 4.3

(a) A red-bellied woodpecker with its tongue extended. (b) The tongues of woodpeckers have rear-facing bristles to help capture and hold insects. [Both Alamy.]

(a)

(b)

FIGURE 4.4

An acorn woodpecker with its granary of acorns. [Alamy.]

With all this pecking, how do woodpeckers avoid brain injury? For a long time, people thought that woodpeckers had a special, foam-like material between their brains and their skulls, a little like the padding in a bicycle helmet. Because the mechanics of foams is my research specialty, when I heard about this, I had to find out more. I discovered that in the 1970s a group of neurologists interested in human brain injury had dissected the head of a pale-billed woodpecker—only to find that there was no special foamy material protecting the brain.

If that's the case, then how do woodpeckers avoid brain injury? I was now hooked and kept investigating.

The same group of neurologists, wanting to know what decelerations and forces woodpecker brains were subjected to, took high-speed videos of an acorn woodpecker pecking into a tree trunk. This was no mean feat back in the 1970s, when filming meant using actual film.

The first challenge for the neurologists was figuring out how to avoid using vast amounts of film, given that they were planning to use film rates of up to 2,000 frames per second. The key was finding a way to trigger the filming just as the pecking started—and they found one when they heard about a park ranger in Placerita Canyon State Park, in Newhall, California. The ranger had an injured acorn woodpecker in the office that would peck in response to the sound of the office IBM Selectric typewriter (which sounds remarkably like an acorn woodpecker pecking). With that trick up their sleeves, the neurologists were able to videotape the acorn woodpecker pecking and, from the position of its head and the elapsed time, calculate the speed at which it was going and the deceleration on impact.

What they found was astonishing: the decelerations the woodpecker was sustaining without brain injury were between 600 and 1,500 g. A more recent study on three different species of woodpeckers measured maximum decelerations of 300–450 g. (1 g is the acceleration an object would undergo from gravity; for instance, a dropped ball accelerates at 1 g.) Human brains, by comparison, typically suffer injury starting around 100 g. Not only that, woodpeckers peck over and over, up to 20 times a second. How do they do it?

It's largely a matter of size. The brain of an acorn woodpecker has a mass of 2 grams, compared with a human brain mass of 1,400 grams (about 3 pounds). Force is equal to mass times acceleration (or deceleration, if stopping), which means that for the same deceleration on impact, the force on a woodpecker brain is much, much smaller than it is on a human brain. For my complete calculations, see box 4.1.

BOX 4.1: WOODPECKER PECKING

Comparison of tolerable deceleration in woodpeckers and humans

When it comes to head-banging and brain damage, what really matters isn't just the force involved. It's the force divided by the area over which the force acts, or what engineers call stress. One way to think about this is to imagine a one-inch-diameter rope that breaks at a load of 1,000 pounds. If there were two identical one-inch-diameter ropes, held in parallel, they would break at a load of 2,000 pounds. Doubling the cross-sectional area doubles the failure load: the stress at failure, or the strength of the ropes, is the same.

The force on the brain is equal to the mass of the brain times the deceleration that it undergoes on impact:

$$force = mass \times deceleration$$

The mass depends on the radius of the brain cubed (assuming it is roughly hemispherical). The area over which the force acts on the brain depends on the radius of the brain squared, so that:

$$stress = \frac{mass \cdot deceleration}{area} = Constant \cdot \frac{density \cdot radius^3 \cdot deceleration}{radius^2}$$

Assuming that the woodpecker brain and the human brain are made of the same tissue, with the same density, and can tolerate the same stress before injury, we can then say that the product of the radius of the brain and the tolerable acceleration is the same in both woodpecker and human brains:

$$radius_{woodpecker} \cdot deceleration_{woodpecker} = radius_{human} \cdot deceleration_{human}$$

Rearranging, we find that the acceleration the woodpecker brain can tolerate is equal to the acceleration the human brain can tolerate times the ratio of the radii of human to woodpecker brains:

$$deceleration_{woodpecker} = deceleration_{human} \cdot \frac{radius_{human}}{radius_{woodpecker}}$$

The radius of a human brain is about eight times that of an acorn woodpecker brain, which means that an acorn woodpecker brain can tolerate eight times the deceleration that a human brain can.

In addition, the hemispherical woodpecker brain is rotated 90° within the skull compared to the human brain, so that the area of contact between the brain and the skull is doubled for the woodpecker. This doubles the acceleration it can tolerate.

The neurologists' high-speed video also revealed that the duration of impact was much less for the woodpecker during pecking than in a typical human brain impact (0.5–1.0 milliseconds vs. 3–15 milliseconds). Brain tissue has been shown to tolerate higher decelerations if they are for a shorter timespan; accounting for this increases the deceleration that the woodpecker brain can withstand by a factor of about three.

Combining all of these factors, the acorn woodpecker brain can tolerate 8 × 2 × 3 = 48 times the deceleration that human brains can tolerate, or about 4,800 g, well above the measured decelerations on the acorn woodpecker of 600–1,500 g.

If you want to learn even more, I suggest that you watch *Built to Peck*, a 25-minute, eight-part video series that I made in conjunction with MITx, MIT's online-education department. The series describes the adaptations of the woodpecker for pecking, how the neurologists measured the deceleration of the acorn woodpecker while pecking, and the mechanics of how woodpeckers avoid brain injury. The first two segments include a brief tour of the bird collection at the Museum of Comparative Zoology of Harvard University and are worth watching just for that. Here's the link: https://www.youtube.com/watch?v=3pnCVwQRxBQ

HUMMINGBIRDS: TONGUES FOR COLLECTING NECTAR

I'm at the Curi-Cancha Reserve in the cloud forest in the mountains near Monteverde, Costa Rica, watching dozens of hummingbirds, their iridescent feathers flashing, their wings humming, as they zip around several feeders, darting in and hovering at an opening in the feeder, probing it with their tongues, and flying off again. It's mesmerizing to see so many hummingbirds at once. I've never seen anything like it. I especially love the violet sabrewing, with its curved bill, tail tipped in white, and its entire body awash in iridescent violets, blues, and greens.

Hummingbirds feed mostly on nectar. The lengths of their bills reflect the lengths of the flowers from which they gather nectar: the longer the bill, the longer the flower that the hummingbird probes (figure 4.5). And there's a twist to this: in some species, such as the hermit hummingbirds, females have more highly curved bills than males and so can gather nectar from more highly curved flowers (figure 4.6).

For nearly 180 years, it was thought that nectar moves up the tubular hummingbird tongue by capillary action, similar to the way water in a narrow, vertical glass tube rises driven by surface tension (figure 4.7). However, a more detailed examination of the bill and tongue by Alejandro Rico-Guevara and Margaret Rubega, at the University of Connecticut, using high-speed video, has shown that the hummingbird tongue doesn't actually work as a capillary tube. Instead, it acts as a tiny pump, trapping nectar and drawing it back into the mouth. This mechanism arises from a combination of the springiness of the tongue itself and the surface tension of the nectar.

Here's how the mechanism works. The hummingbird tongue has an unusual structure (figure 4.8). Toward the tip, it is split longitudinally into two separate halves, each with a rod running along its length. Over the last quarter-inch (6 mm) of the tongue, tiny lateral fringes, or lamellae, curve away from the rods; the fringes are thicker toward the base of the bill and thinner and more delicate toward the tip. For most of the length of the rods, the fringes are below the rod, but at the very outer tip of the tongue, the rods bend downward, and the fringes are above the rods.

As the tongue passes through the bill to extend into the flower, the bill opens only just enough for the tongue to pass through, compressing it, so that the cross section of the two rods and their corresponding fringes looks like two flat ovals (figure 4.9, top row). As the tongue starts to move out of the bill, a thin layer of nectar remains on the tongue from the previous lick into the nectar pool within

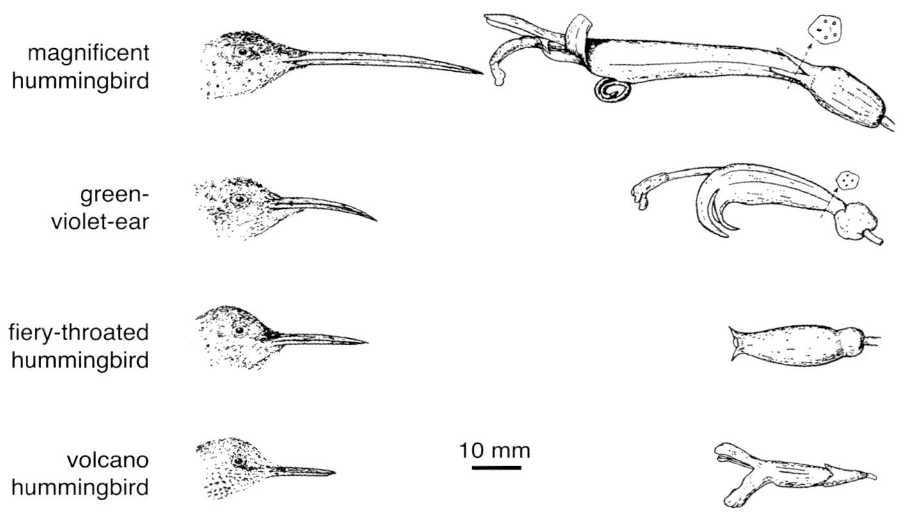

FIGURE 4.5

The lengths of hummingbirds' bills reflect the lengths of the flowers from which they gather nectar. [From Wolf et al. (1976), reproduced with permission.]

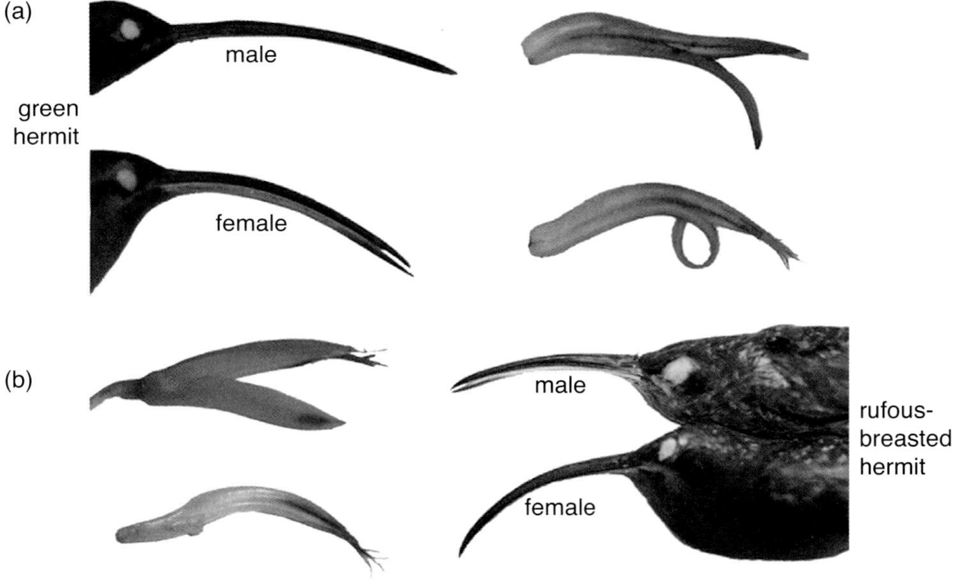

FIGURE 4.6

Female hermit hummingbirds have more highly curved bills than males and so can gather nectar from more highly curved flowers. [From Temeles et al. (2010), reproduced with permission.]

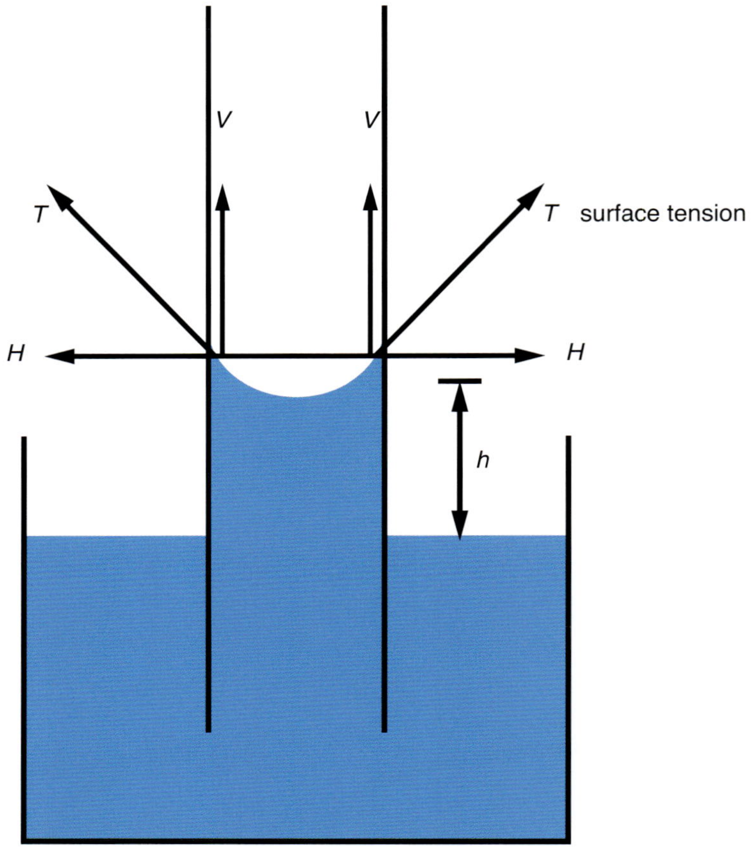

FIGURE 4.7

Liquid drawn up a capillary tube by surface tension.

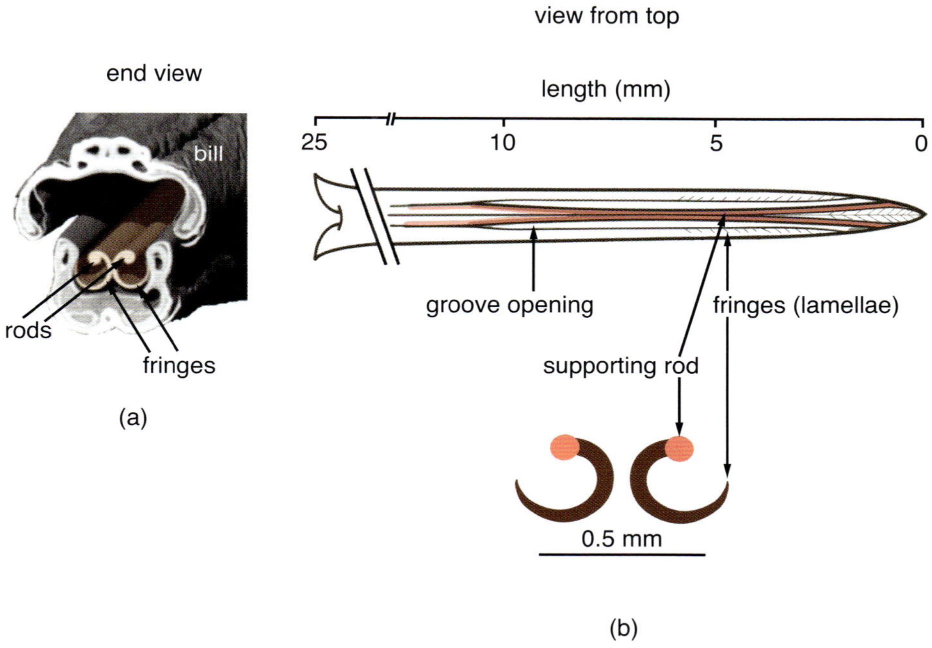

FIGURE 4.8

The structure of hummingbirds' tongues: (a) cross section, (b) view from top. The end of the tongue is split into two, each half with a stiff rod running along the length. At the tip, fringes attach to the rods. [(a) From Rico-Guevara et al. (2015), reproduced with permission; (b) after Rico-Guevara and Rubega (2011).]

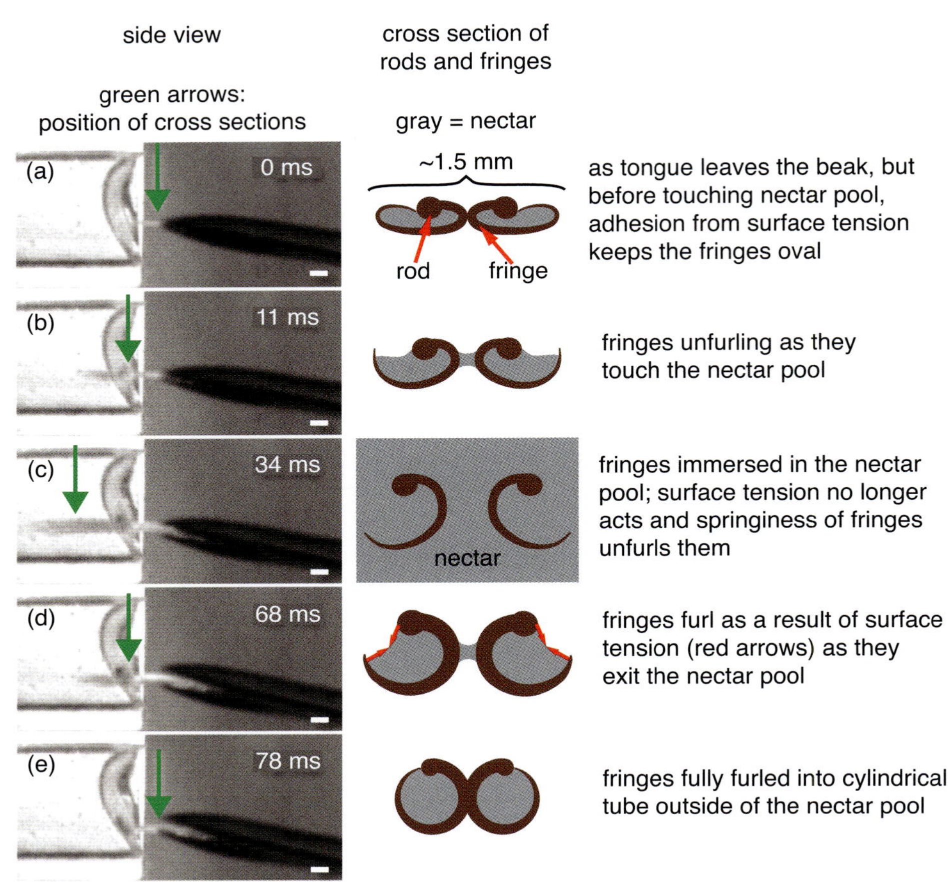

FIGURE 4.9

Left column: photographs from high-speed video of a single lick of a hummingbird tongue as it enters and leaves a glass tube of nectar (ms = milliseconds = 0.001 seconds). Right column: cross sections of the rods and fringes. [From Rico-Guevara and Rubega (2011), reproduced with permission.]

the flower; this is shown in gray within the fringes in the slide. This thin layer of nectar within the fringes gives rise to adhesive forces from surface tension (like the force you feel when pulling two wet sheets of plastic apart) that keep the fringes in the compressed oval shape until the tip of the tongue touches the pool of nectar in the flower. At this point (figure 4.9, third row), the tip of the tongue is surrounded by nectar—and the surface tension, which has been holding each half of the end of the tongue in the compressed oval shape, disappears (because surface tension occurs only at a surface). The fringes at the tip unfurl, driven by the springiness of the fringe material (similar to the expansion of a compressed spring when the compressive force is removed). As the tongue extends further into the nectar pool, more and more of the fringes along the length of the tongue unfurl. The unfurling of the fringes at the interface between the air and the pool of nectar is visible in the photograph in figure 4.10 and is shown schematically in figure 4.11.

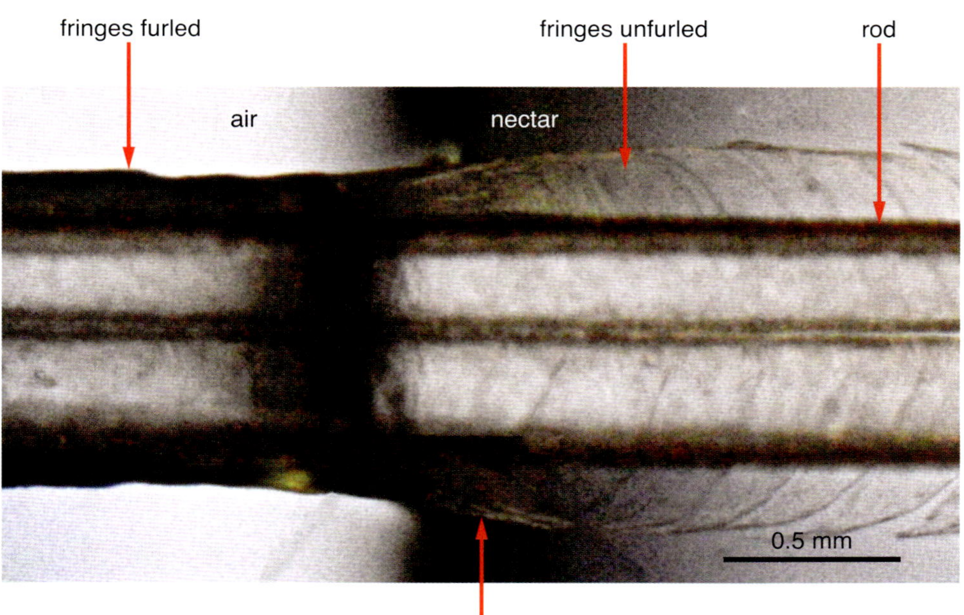

FIGURE 4.10

Unfurling of the fringes as they enter the nectar in the glass tube. [Adapted from Rico-Guevara and Rubega (2011), reproduced with permission.]

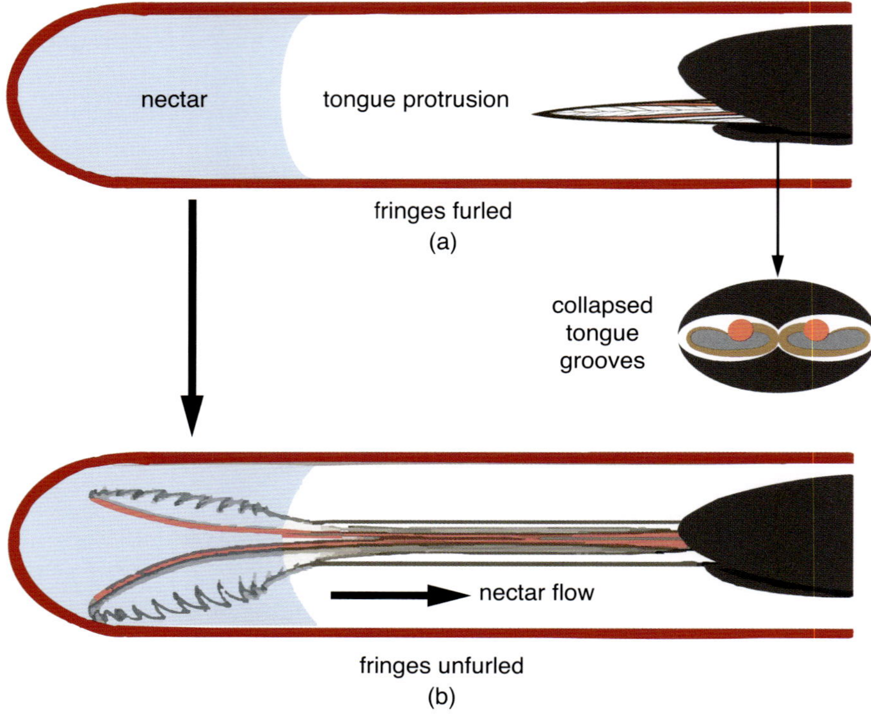

FIGURE 4.11

Schematic showing the unfurling of the fringes on the hummingbird tongue in the nectar. [After Rico-Guevara et al. (2015), with permission.]

Conversely, as the tongue withdraws from the pool of nectar in the flower, the fringes furl up again, driven by surface tension forces, forming cylindrical tubes to convey the nectar to the bird's mouth (figure 4.9, bottom two rows).

Who would have thought that a hummingbird draws nectar into its mouth using a pump action created by the interplay between the surface tension of the nectar and the springiness of the fringes on its tongue? A wonderful discovery.

CALIDRIS SANDPIPERS: BILLS FOR SENSING PREY

Shorebirds are often seen probing the wet sand at the water's edge with their bills. Some, like semipalmated plovers, find food on the surface, by sight. Others, like American oystercatchers, randomly probe into the sand until they detect prey by touching it with their bills. *Calidris* sandpipers, such as sanderlings, western sandpipers, and red knots, too, probe into the sand with their bills, often with a rapid up-and-down "stitching" motion, like that of a sewing-machine needle. Their straight, thin, relatively short bills even look a little like needles. Watching them, I had always thought that they were just randomly hoping to hit something to eat, but they're not. Remarkably, they can detect prey beneath the surface of the sand without directly touching it. How do they do it?

One early study—conducted in 1986 by AFC Gerritsen and A Meiboom, at the Rijksuniversiteit Leiden, in the Netherlands—focused on sanderlings (*Calidris alba*), which prey on annelid worms (*Nerine cirratulus*) that are about an inch or two long (25–50 mm) and burrow in wet sand up to a foot (300 mm) beneath the surface, leaving visible tracks. Gerritsen and Mieboom placed *Nerine cirratulus* worms in the bottom half of trays filled about 1.5 inches (40 mm) deep with wet sand. Individual trays had varying numbers of worms and varying numbers of simulated worm tracks on the surface.

In trays with no prey, the sanderlings pecked at the surface, but didn't probe below. In trays with visible worm tracks, the birds probed near the tracks, but continued probing for less than a minute if there were no worms in the tray. In trays with high numbers of prey but no tracks, however, the birds probed in a pattern that followed the placement of the worms, and once one was caught, the birds caught others nearby. Somehow, the birds were sensing the worms beneath the surface without touching them.

Needless to say, they were doing it with their bills—more specifically, with sensory cells in the tips of their bills, known as Herbst corpuscles, that can detect minute changes in pressure. Bills are mostly bone overlaid by a thin outer sheath of keratin. The sensory cells in *Calidris* sandpipers are located in small pits within the bone of the bill, just beneath the outer keratin layer. Figure 4.12 shows the pits in the bills of dunlin (*Calidris alpina*). When worms move in wet sand, they create tiny waves in the water between the sand grains, and these waves give rise to small changes in water pressure. When a *Calidris* sandpiper probes its bill into the sand, the Herbst corpuscles in its bill detect these pressure changes, allowing the bird to detect prey without touching it; Gerritsen and Meiboom called this ability "remote touch." It's estimated that sanderlings can detect prey up to an inch (25 mm) from their bills.

Can *Calidris* sandpipers detect prey that doesn't move? The red knot can. Most of the year, when not breeding, red knots eat mollusks and hard-shelled crustaceans that don't move much, and they locate them by probing in intertidal zones. Their bills also have sensory pits. So do they use remote touch?

To answer this question, in 1998 Theunis Piersma and his colleagues in the Netherlands designed experiments with red knots (*Calidris canutus*) and live Baltic clams (*Macoma balthica*) that are about 25 mm (1 inch) long and don't move. The birds were presented with three buckets: one that contained only wet sand, another than contained wet sand and clams, and a third that contained wet sand and clam-sized stones (figure 4.13). In the latter two buckets, the clams and stones were placed lower than the length of the red knot's bill. Fascinatingly, the birds avoided the buckets containing only sand and instead preferentially probed those containing clams and stones, even though they could not touch them. The birds probed the buckets with stones the same amount as the buckets with clams, confirming that they were not using smell, taste, vibrational signals, temperature gradients, or electromagnetic fields to detect the clams. Red knots were indeed using remote touch to locate stationary prey.

In a variation on the experiment, the researchers placed either the clams or stones in a bucket with dry sand and also made available a bucket of wet sand with no objects. The birds probed all the buckets equally, suggesting that the sand has to be wet for their sensory system to work. It turns out that when the bird's bill repeatedly probes wet sand, with no other objects in it, the act of probing creates tiny waves in the water around the bill that move symmetrically away from the bill. But if there is a rigid object in the wet sand, such as a clam or stone, then the wave pattern

(a)

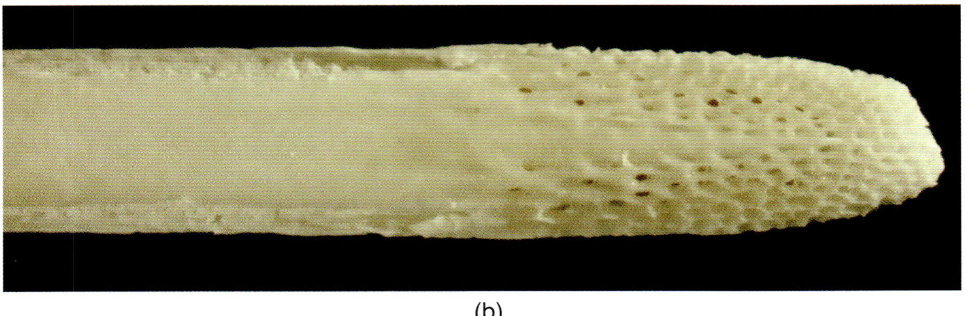
(b)

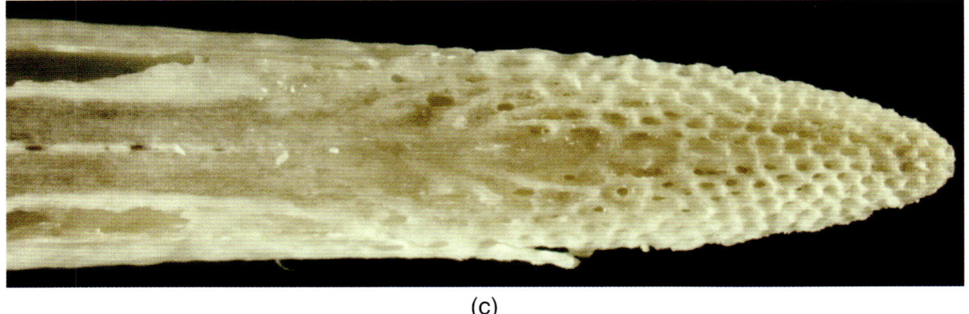
(c)

FIGURE 4.12

(a) A dunlin (*Calidris alpina*). (b, c) The tips of a dunlin's bill have pits in the bone (shown here) with pressure-sensing Herbst's corpuscles (not shown). Images of bills with skin removed and bleached: (b) female dunlin, upper bill, (c) male dunlin, lower bill. [(a) Alamy; (b, c) unpublished photos courtesy Silke Nebel, Dan Jackson, and Bob Elner.]

FIGURE 4.13

Red knots peck at buckets of wet sand containing either mollusks or stones, placed below the depth that their bills can reach, in preference to buckets of just wet sand. They, too, have pressure-sensing cells in their bills. [Top: Alamy.]

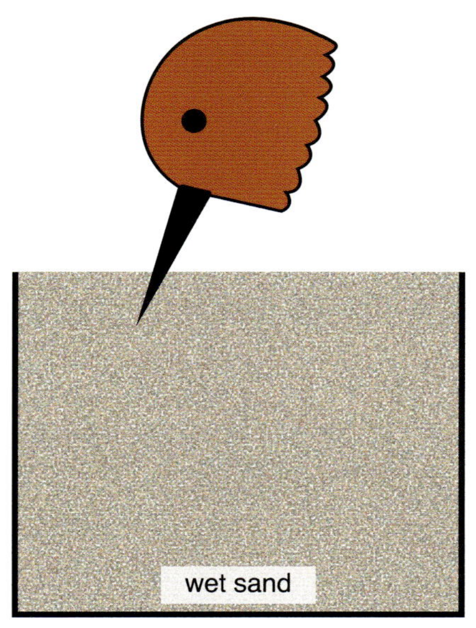

is disturbed (for instance, reflected off the object and back to the bill), causing a change in pressure at the bill that the sensory cells detect.

The rapid "stitching" motions that *Calidris* sandpipers make in wet sand produce a series of waves in the water between the sand grains. If there is an object nearby, the waves will be reflected multiple times, stimulating the pressure cells in their bills repeatedly, making it easier to detect the object than if they probed only once.

PHALAROPES: BILLS FOR CAPTURING PLANKTON

It's summer, and I'm at the beach on Cape Cod, where I spot a Wilson's phalarope in the water, spinning around in tight circles. I'm excited to see it, as it's an uncommon bird at the Cape, but I'm a bit baffled by the spinning. Later I learn that it was deliberately creating a vortex in the water to pull plankton toward it to eat.

Plankton are organisms that drift with water currents, unable to swim against them. (Their name comes from the Greek word *planktos*, meaning "wanderer" or "drifter.") Plankton are typically small, which makes it challenging to pluck them out of the water—think of trying to pluck a small particle out of a glass of water with your fingers. Filter feeders, such as flamingos and northern shovelers, capture plankton by flushing water through their bills, trapping them in the comb-like projections at the edges of their bills or tongues. But phalaropes aren't filter feeders and don't have those comb-like projections. And they don't have lips to suck water up into their mouths. So how do they get water droplets filled with plankton into their mouths?

Phalaropes capture plankton by pecking at the water surface and trapping droplets containing plankton in their open bills. Once a droplet is in the bill, phalaropes rapidly and repeatedly open and close it slightly. High-speed video, taken by Margaret Rubega and Bryan Obst in 1993, when they were at the University of California, Irvine, has revealed that as phalaropes do this, the plankton-containing droplet progressively ratchets up the bill with each cycle of opening and closing (figure 4.14). As the bill is opened and closed, the droplet is pulled and squeezed, changing its shape in such a way that surface tension forces at the outer edge of the drop draw the droplet along the beak, toward the mouth. From the time a droplet is pecked out of the water, it typically takes 0.02 seconds or less for a phalarope to move the droplet along its beak and into its mouth, which allows it to consume large numbers of prey efficiently. (To eliminate the possibility that the tongue is somehow

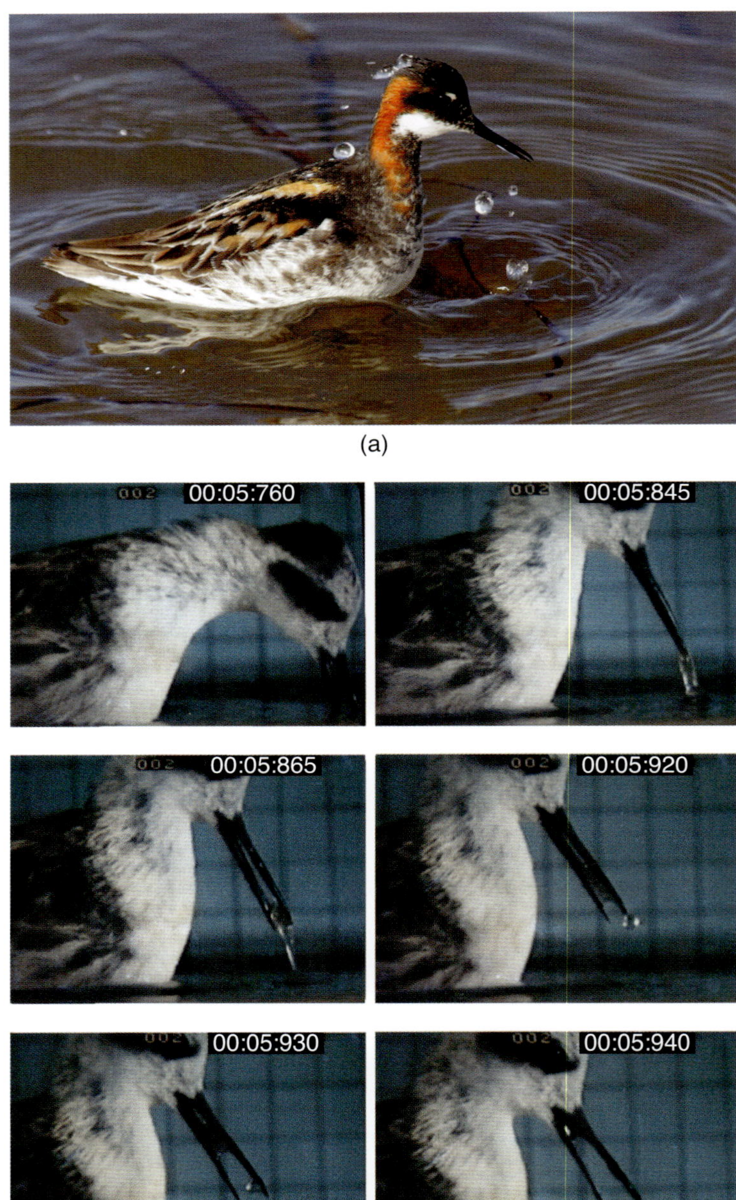

FIGURE 4.14

Bills for capturing plankton. (a) A red-necked phalarope spinning in the water, producing a vortex to draw plankton toward it. (b) Still photos from high-speed video of a red-necked phalarope repeatedly opening and closing its bill to draw a drop of water containing plankton to its mouth. Numbers on each still photo are time stamps; e.g., at top left, 5.760 seconds. [(a) Alamy; (b) from Rubega and Obst (1993), Reproduced by permission of Oxford University Press on behalf of the American Ornithological Society.]

used to move the water droplet toward the mouth, Rubega and Obst reproduced the droplet motion in the bill of a dead phalarope.)

This "surface tension transport" of prey-containing water droplets, first observed in the red-necked phalarope, has now been observed in Wilson's phalarope as well as a variety of small and medium-sized shorebirds with needle-like bills that feed standing in shallow water: the western sandpiper, the least sandpiper, the little stint, the sanderling, the dunlin, the curlew sandpiper, the common redshank, and the black-winged stilt (figure 4.15). All have bills that are relatively straight, less than 3 inches (75 mm) long, and less than a quarter-inch (6 mm) wide (figure 4.16). The width of the bill affects the shape of the drop and the surface over which surface tension acts. In addition, a water drop within an open bill has an asymmetric shape that also contributes to differing surface tension forces at the front and back of the drop (figure 4.16, inset at top). As the bill opens and closes, the shape of the droplet changes; surface tension forces (which, remember, act tangent to the surface of the droplet) incrementally move the drop toward the mouth with each cycle of opening and closing of the bill. It is the shape of their bills that allows phalaropes and other shorebirds to ratchet water droplets up their bills and into their mouths.

FIGURE 4.15

Birds that use "surface tension transport" to move water droplets containing plankton up their bills. All have bills that are relatively straight, less than 3 inches (75 mm) long, and less than a quarter inch (6 mm) wide. [Photos all Alamy.]

Wilson's phalarope

western sandpiper

least sandpiper

little stint

sanderling

dunlin

curlew sandpiper

common redshank

black-winged stilt

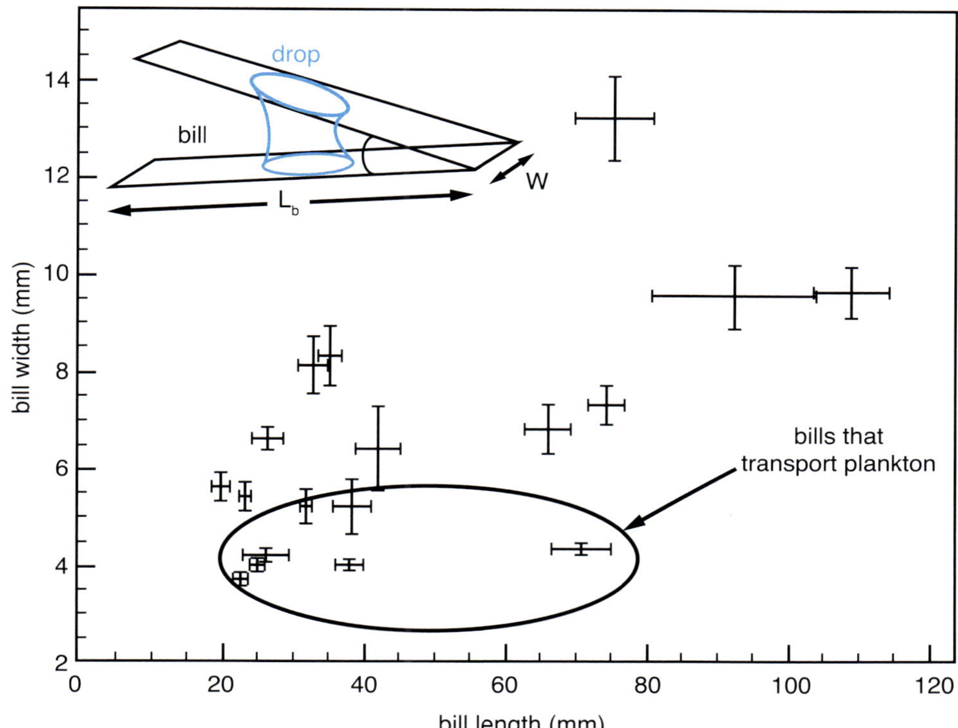

FIGURE 4.16

Bills for capturing water droplets containing plankton. Birds in the oval envelope are red-necked phalarope, Wilson's phalarope, black-winged stilt, dunlin, little stint, sanderling, and western sandpiper. All have bills that are relatively straight, less than 3 inches (75 mm) long, and less than a quarter-inch (6 mm) wide. (Birds outside the envelope are common snipe, Eurasian oystercatcher, Kentish plover, common ringed plover, spotted redshank, black-tailed godwit, gray plover, northern lapwing, ruff, and ruddy turnstone.) [After Prakash et al. (2008).]

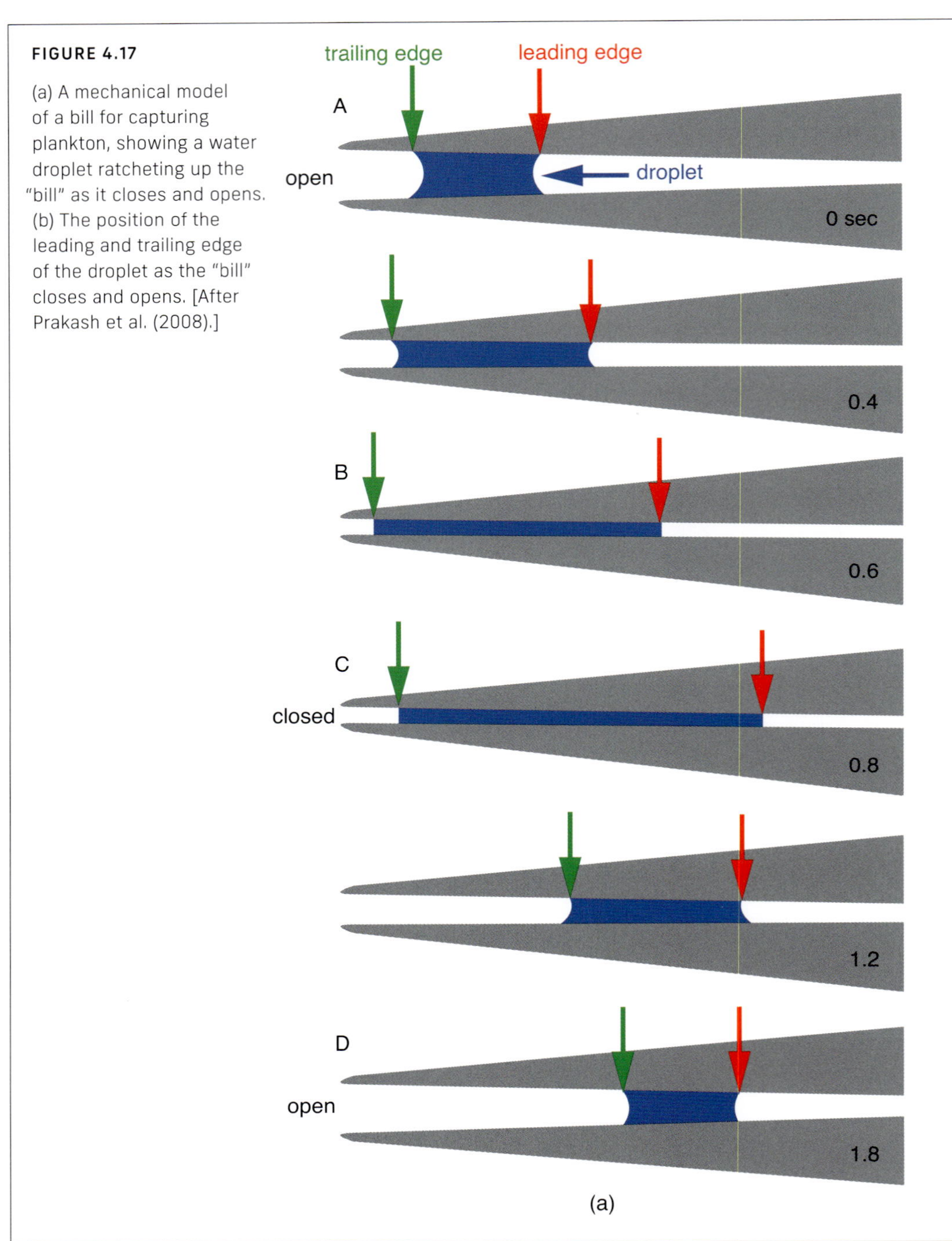

FIGURE 4.17

(a) A mechanical model of a bill for capturing plankton, showing a water droplet ratcheting up the "bill" as it closes and opens. (b) The position of the leading and trailing edge of the droplet as the "bill" closes and opens. [After Prakash et al. (2008).]

BOX 4.2: A MECHANICAL MODEL OF A BILL

A group of scientists at MIT and at the Ecole Supérieure de Physique et de Chimie Industrielles in Paris, led by John Bush, at MIT, have reproduced surface-tension-driven droplet motion in a stainless-steel replica of a bill that was opened and closed by a motorized, computer-controlled fixture. Using high-speed video, the scientists captured the motion of the water droplet along the mechanical bill. Looking at the images in figure 4.17a, we can see that as the bill closes (images A, B, and C, at 0.4, 0.6, and 0.8 seconds), the trailing edge of the drop (green arrows) remains more or less still while the leading edge (red arrows) moves toward the mouth. Then, when the bill starts to reopen (at 1.2 seconds and at D, 1.8 seconds), the trailing edge of the drop scoots further toward the mouth while the leading edge stays more or less in the same place. The alternating ratcheting of the leading and trailing edges as the bill closes and opens moves the droplet toward the mouth. The measurements of the positions of the leading and trailing edges of the water droplet as the bill opens and closes are shown in figure 4.17b. As the bill closes and opens, the droplet changes shape and the surface tension forces acting on the surface of the droplet change, too; Bush and his team have analyzed the details of how the changes in surface tension ratchet the drop forward.

As I was preparing to give a talk on bird bills, I contacted Bush to see if I could take a photograph of his mechanical "bill" for my talk. I was amused by his response: he said he no longer had it, because it was just a pair of mini needle-nosed pliers, readily available from woodworking supply stores, with a geometry similar to that of a bird bill that transports plankton by surface tension.

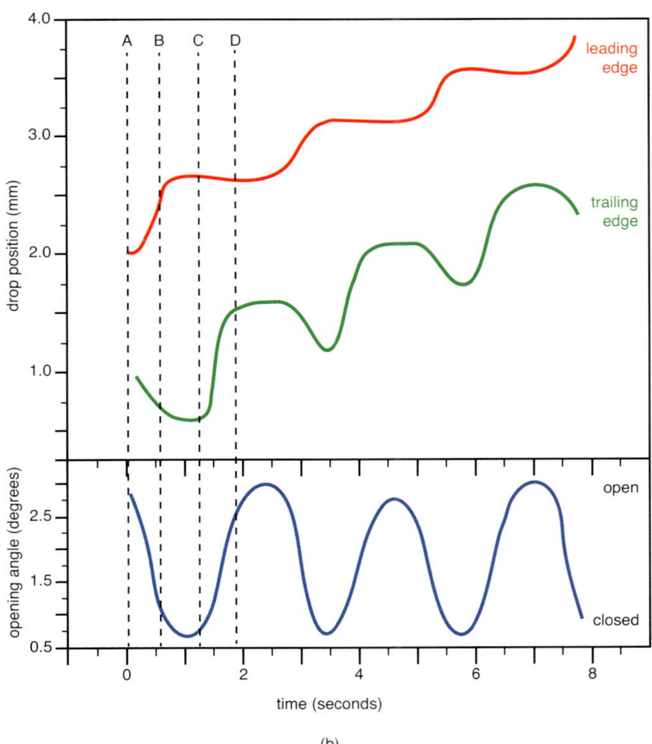

(b)

Now, when I go birding, I don't just see birds' bills from the outside. In my mind's eye, I see what's going on inside bills that enables them to find and capture their prey. When I see a hairy woodpecker at Jamaica Pond pecking on a decaying branch, I think of its hyoid contracting to extend its tongue deep into the hole to capture a bug. When a ruby-throated hummingbird is hovering at the pink honeysuckle flowers in my back garden, I imagine its double-barreled tongue, fringes opening as they enter the nectar and closing as they withdraw from it. In Provincetown, seeing *Calidris* sandpipers needle-nosing on the beach, I'm thinking of their pressure-sensing cells detecting worms and mollusks without actually touching them. And on the rare occasion that I see a phalarope twirling in the water and drawing droplets into its mouth, I think of Bush and his needle-nose-plier experiments that showed in detail how surface tension forces pull the droplet into its mouth. The hidden secrets of bills are amazing.

REFERENCES

INTRODUCTION

Ehrlich PR, Dobkin DS, and Wheye D (1988). *The Birders Handbook*. Simon and Schuster.

Gill FB (2007). *Ornithology*. Third edition. WH Freeman.

Lederer RJ (2016). *Beaks, Bones and Bird Songs: How the Struggle for Survival Has Shaped Birds and Their Behavior*. Timber Press.

Lovette IJ and Fitzpatrick JW, eds. (2016). *The Cornell Lab of Ornithology Handbook of Bird Biology*. Third edition. Wiley.

Matsui H, Hunt GR, Oberhofer K, Ogihara N, McGowan KJ, Mithraratne K, Yamasaki T, Gray RD, and Izawa EI (2016). Adaptive bill morphology for enhanced tool manipulation in New Caledonian crows. *Scientific Reports* **6**, 22776.

Proctor NS and Lynch PJ (1993). *Manual of Ornithology: Avian Structure and Function*. Yale University Press.

Rico-Guevara A, Rubega MA, Hurme KJ, and Dudley R (2019). Shifting paradigms in the mechanics of nectar extraction and hummingbird bill morphology. *Integrative Organismal Biology* **1**, 1–15.

Rutz C, Sugasawa S, van der Wal JEM, Klump BC, and St Clair JJH (2016). Tool bending in New Caledonian crows. *Proc. Roy. Soc. Open Science* **3**, 160439.

Tattersall GJ, Andrade DV, and Abe AS (2009). Heat exchange from the toucan bill reveals a controllable vascular thermal radiator. *Science* **325**, 468–470.

Weiner J (1994). *The Beak of the Finch: A Story of Evolution in Our Time*. Alfred A. Knopf.

WOODPECKERS: BILLS FOR PECKING

Built to Peck (n.d.). Video series. https://www.youtube.com/watch?v=3pnCVwQRxBQ&list=PLTz-v-F773kXzGHaX1-zV6zoMPCxV8GJM

Gibson LJ (2006). Woodpecker pecking: how woodpeckers avoid brain injury. *J. Zoology* **270**, 462–465.

May PRA, Fuster JM, Haber MS, and Hirschman A (1979). Woodpecker drilling behavior: an endorsement of the rotational theory of impact brain injury. *Arch. Neurol.* **36**, 370–373.

May PRA, Fuster JM, Newman P, and Hirschman A (1976). Woodpeckers and head injury. *Lancet* **1**, 454–455.

NASA (n.d.). STS-70 [space shuttle and northern flickers]. https://www.nasa.gov/mission/sts-70/

Sawyer, K (1995). NASA's scare tactics keep birds at bay. *Washington Post*, July 11.

Van Wassenbergh S, Orlieb EJ, Mielke M, Boehmer C, Shadwick RE, and Abourachid A (2022). Woodpeckers minimize cranial absorption of shocks. *Current Biology* **32**, 3189.

HUMMINGBIRDS: TONGUES FOR COLLECTING NECTAR

Rico-Guevara A, Fan T-H, and Rubega MA (2015). Hummingbird tongues are elastic micropumps. *Proc. Roy. Soc.* **B282**, 2015.1014.

Rico-Guevara A and Rubega MA (2011). The hummingbird tongue is a fluid trap, not a capillary tube. *PNAS* **108**, 9356–9360.

Temeles EJ, Miller JS, and Rifkin JL (2010). Evolution of sexual dimorphism in bill size and shape of hermit hummingbirds (*Phaethornithinae*): a role for ecological causation. *Phil. Trans. Roy. Soc.* **B365**, 1053–1063.

Wolf LL, Stiles FG, and Hainsworth FR (1976). Ecological organization of a tropical, highland hummingbird community. *J. Anim. Ecology* **45**, 349–379.

CALIDRIS SANDPIPERS: BILLS FOR SENSING PREY

Gerritsen AFC and Meiboom A (1986). The role of touch in prey density estimation by *Calidris alba*. *Netherlands Journal of Zoology* **36**, 530–562.

Nebel S, Jackson DL, and Elner RW (2005). Functional association of bill morphology and foraging behaviour in calidrid sandpipers. *Animal Biology* **55**, 235–243.

Piersma T, van Aelst R, Kurk K, Berkhoudt H, and Maas LRM (1998). A new pressure sensory mechanism for prey detection in birds: the use of principles of seabed dynamics? *Proc. R. Soc. Lond.* **B265**, 1377–1383.

PHALAROPES: BILLS FOR CAPTURING PLANKTON

Bush JWM, Peaudecerf F, Prakash M, and Quéré D (2010). On a tweezer for droplets. *Advances in Colloid and Interface Science* **161**, 10–14.

Estrella SM, Masero JA, and Perez-Hurtado A (2007). Small-prey profitability: field analysis of shorebirds' use of surface tension of water to transport prey. *Auk* **124**, 1244–1253.

Prakash M, Quéré D, and Bush JWM (2008). Surface tension transport of prey by feeding shorebirds: the capillary ratchet. *Science* **320**, 931–934.

Rubega MA (1996). Surface tension prey transport in shorebirds: how widespread is it? *Ibis* **139**, 488–493.

Rubega MA and Obst BS (1993). Surface-tension feeding in phalaropes: discovery of a novel feeding mechanism. *Auk* **110** (2), 169–178.

5 EGGS

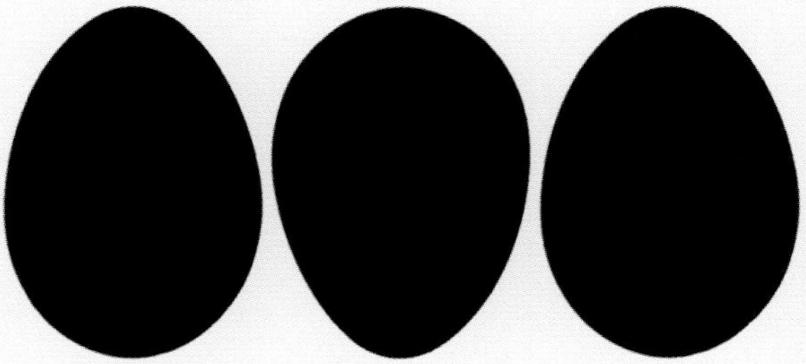

I think, that, if required, on pain of death, to name instantly the most perfect thing in the universe, I should risk my fate on a bird's egg.

Thomas Wentworth Higginson

I'm at Harvard's Museum of Comparative Zoology to look at the egg collection in the Ornithology Department. Kate Eldridge, a curatorial assistant, takes me into the collection room where the entire back wall is lined with a row of red cabinets filled with drawer after drawer of eggs, mostly collected over a century ago. As she opens a cabinet and pulls out a drawer, I see an array of small boxes, each with a few eggs, and each with a label listing the species, date, location collected, and the name of the collector, many hand-written in ink on small, yellowed pieces of paper, in a Victorian style of script that looks like it belongs in a British drama from Masterpiece Theatre. Most of the boxes contain just the eggs, but some have the entire nest with its eggs and even the branches holding the nest. I'm captivated just looking at it all.

I can't get over the sheer variety (figure 5.1). Colors ranging from white, cream, buff, brown, and rust to blues and even greens, some with dark brown speckles or Jackson Pollock-style drizzles. Glossy tinamou eggs, shiny as a billiard ball, in solid

FIGURE 5.1

Bird eggs of a variety of colors and shapes. [Museum of Comparative Zoology, Harvard University, specimens 352849, 367177, 358288, 354349, 355101, 353571, 361778, 354278, 352291, 353409, © President and Fellows of Harvard College.]

colors of milk chocolate brown and avocado green. Tiny hummingbird eggs, half the size of a grape, huge ostrich eggs, the size of a small cantaloupe. Spherical owl eggs, elliptical swan eggs, asymmetric murre eggs, blunt at one end, pointy at the other. What accounts for all the different colors, sizes, and shapes of eggs? And, more fundamentally, how are eggs even produced?

OVUM TO EGG

When I sat down to write this chapter, I was a little embarrassed to realize that I didn't actually know how, in birds, an egg cell transforms into the egg that we're all familiar with. I assumed that the egg cell had to be fertilized by sperm, but what happens after that? The materials scientist in me wanted to know what the process is that gives the structure of an egg: the *yolk*, which supplies food and energy to the embryo until it develops into a chick ready to hatch; the *albumen*, or egg white, the clear, viscous fluid surrounding the yolk that provides water and acts as a shock absorber, cushioning the embryo against impacts; and the *shell*, which is hard enough to protect the developing chick but just porous enough to allow oxygen, carbon dioxide, and water vapor, all necessary to maintain life, to flow in and out of the egg.

The process starts in the ovary, with yolk being deposited around an egg cell, or ovum. Each ovum sits within a follicle which extracts the components of the yolk from the bloodstream and deposits them around the ovum, in alternating layers of yellow globules that contain mostly fat and pigment, deposited during the day, and white globules that contain mostly protein, deposited at night (figure 5.2). In most birds, the yolk is deposited over four to eight days, but it can take up to two weeks in some species. The yolk contains the nutrients, hormones, and immune factors that nourish and sustain the developing embryo once the ovum is fertilized. Yolks are deposited around sufficient ova (plural of ovum) for a clutch of eggs before any are released from the ovary.

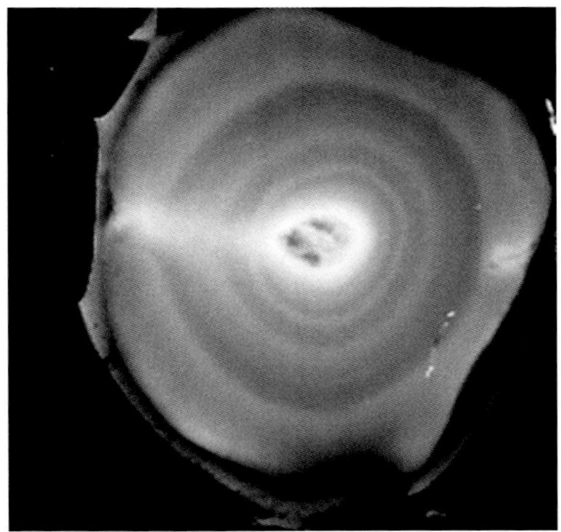

FIGURE 5.2

Alternating dark bands of fat and pigment deposited in the yolk during the day, and light bands of protein deposited at night. [Photograph courtesy of David W Winkler.]

When a female bird ovulates, the ovum is released from the ovary and enters a serpentine tube called the oviduct (figure 5.3). For the next day or two, as the ovum passes through the oviduct, an assembly-line process takes place, and when it's over, what emerges at the other end is an egg. One ovum passes completely through the oviduct and is laid as an egg before the next is released from the ovary, which means that birds carry only one egg at a time, keeping their weight low for flight.

Fertilization by sperm occurs in the funnel-like top end of the oviduct. Fertilization is not necessary for ovulation to occur or an egg to be

FIGURE 5.3

The passage of an ovum through the oviduct to produce an egg. [© Cornell Lab of Ornithology.]

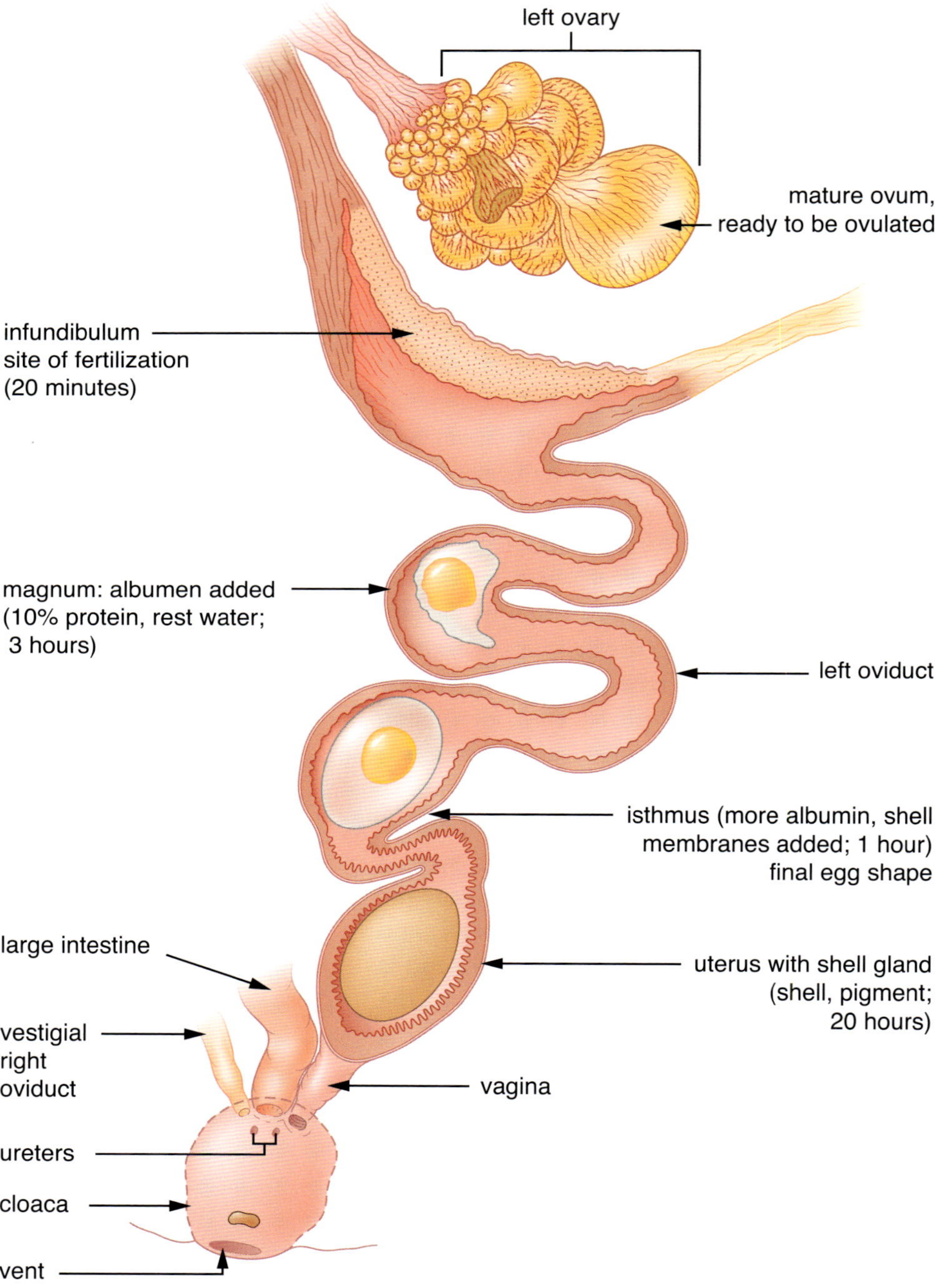

laid. In the wild, there are abundant mating opportunities, which means that nearly all eggs are fertilized, but the chicken eggs that we buy from the supermarket are not.

As the ovum or embryo passes through the bends of the oviduct, it is coated with albumen (made up of 90% water and 10% protein), secreted by glands in the interior walls of the oviduct. At a widening further along the oviduct, the shell membrane is added; this is not the shell itself but rather the membrane underneath it that you can peel off a hard-boiled egg. The shell membrane has two layers, the inner membrane, attached to the albumen, and the outer membrane, to which the shell attaches in the next step of the process. At this stage the "egg" has its final shape. It then passes into the uterus where—continuing the assembly-line process—the shell and pigments are deposited by the shell gland.

The shell gland, on the inside wall of the uterus, acts like a collection of tiny spray nozzles sequentially secreting various materials onto the "egg" to form the shell and add pigments. First, a solution of calcium carbonate (the mineral in chalk) is sprayed onto the membrane, forming foamy blobs that cover the surface of the outer membrane. After those blobs have dried, a second set of nozzles sprays water, which passes between the dried blobs of calcium carbonate and through the shell membrane into the albumen, causing it to swell and nearly fill the membrane. At this point, another set of nozzles sprays another solution of calcium carbonate on top of the dried foam blobs, creating parallel columns of calcite crystals (a form of calcium carbonate). Between groups of calcite-crystal columns, pore channels remain, allowing oxygen, carbon dioxide, and water vapor to pass across the shell (figure 5.4). Once the shell has been deposited, another set of nozzles (pigment glands) sprays the base color onto the shell and, if needed, yet another set of nozzles sprays a pattern of splotches and drizzles. Finally, the outer surface of the egg is coated in a thin, frothy layer of sticky protein that prevents microorganisms from invading the egg.

For a chicken egg, it takes about 24 hours for the ovum or embryo to travel through the oviduct and be laid as an egg; roughly 80% of this time is spent in the uterus. In other species, this process can take up to a week.

During the breeding or laying season, female birds continue to lay eggs until they have a full clutch and begin incubation. That's why domesticated chickens lay eggs more or less continuously, one every day or two for months at a time: if an egg is removed before a hen has a clutch, she will keep laying more. Chickens have been bred to have a particularly long laying season: one chicken, in a remarkably long and productive run of laying, managed to lay 352 eggs in 359 days.

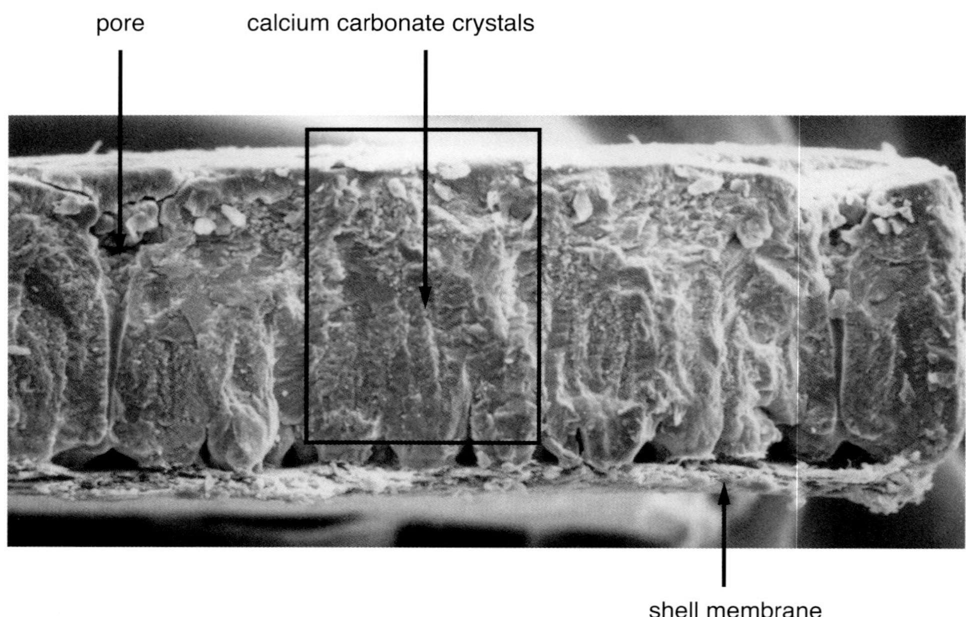

FIGURE 5.4

Scanning electron microscope image through the thickness of an eggshell. [Micrograph courtesy of University of Colorado Museum of Natural History, Paleontology Section, Eggshell Collection, UCM#HEC 563-C, CC BY NC 4.0, with permission.]

Chickens—and other birds that continue to lay eggs if any are removed from the nest before incubation begins—are called "indeterminate layers." But not all species of birds are indeterminate layers. In one study of 46 species of birds, roughly half did not lay additional eggs if one was removed from the nest during the laying period. These birds are called "determinate layers."[1]

The number of eggs in a clutch varies for different species of birds: for instance, ruddy ducks typically have six to ten eggs in a clutch, while bald eagles have only one to three. For a particular species, the size of the clutch is generally larger in breeding seasons when more resources (food, minerals) are available.

COLOR

Looking through drawers of eggs at the Museum of Comparative Zoology, I notice that many are white, while others have the yellowish hue of cream or buff, and yet others are blue or bluish green. In nearly all birds, these colors are produced by just two pigments: protoporphyrin, which creates the creams, buffs, rusts, and browns;

and biliverdin, which creates the blues and greens. Strangely, birds produce blue pigment for their eggs but not their feathers and skin; the reason there are no blue pigments in feathers and skin remains a mystery. (As we saw in chapter 1, the blues in bird feathers are structural colors; so is blue in skin.)

Without added pigment, the shell of an egg is white, the color of calcium carbonate. Eggs that are hidden, such as those laid in cavity nests in trees (by woodpeckers, for example) or in holes in embankments (kingfishers), are white. Birds that begin incubation as soon as the first egg is laid (hawks and owls) or that hide their eggs by covering them when they are not on the nest (grebes) produce white eggs, too.

Birds that nest in the open typically have eggs that are well camouflaged with the ground around them. For instance, many species of killdeer, terns, plovers, and quail have eggs that are buff-colored with darker drizzles, which makes them disappear against the ground on which they are laid (figure 5.5). One study has even

FIGURE 5.5

Little ringed plover eggs on a pebble beach. [Alamy.]

found that female Japanese quail select the nest site that most closely matches their own eggs, suggesting that they know what their own eggs look like.

Ostriches are an exception. They produce large off-white or cream-colored eggs that are conspicuous if neither adult is on the nest, a simple, shallow depression in the ground. But there's a tradeoff at play here: in the open, semiarid-to-desert African plains on which ostriches live, there's nowhere to hide eggs and protect them from the sun and the heat. In that context, white eggs are thought to confer an advantage, because they reflect solar (and possibly ultraviolet) radiation, protecting the eggs when the sun bears down on them and temperatures are high. There is little risk of predation in these areas, since one of the adults always stays near the nest and adult ostriches are generally able to drive off the only predator of the eggs, the Egyptian vulture.

Some birds that make nests that are open at the top, like a cup, such as the American robin, lay pale blue eggs. Studies on domestic poultry have found that embryos develop more quickly if light penetrates the shell. Not only that, daylight hitting a blue eggshell transmits blue light to the embryo, and blue light has been shown to accelerate embryonic growth more than any other color. Blue eggshells thus reduce the time that robin eggs need to develop into chicks and hatch—time during which the eggs are vulnerable to predation. Green is adjacent to blue on the color spectrum (figure 1.3), so bluish green eggs may also enhance early hatching.

Until recently, scientists thought that all the colors of bird eggshells could be produced by just the two pigments mentioned above: protoporphyrin and biliverdin. But Randy Hamchand, Daniel Hanley, Rick Prum, and Christian Bruckner thought that there might be something more to the purplish-brown eggshells of the spotted nothura and the stunning guacamole-green eggshells of the elegant-crested tinamou (figure 5.6). (These are two species of tinamou, a group of mostly ground-dwelling South American birds that look a little like grouse—and to my delight I was able to see examples of their eggs when I visited the Museum of Comparative Zoology.) After investigating the matter, they discovered that the purplish-brown and guacamole-green colors are produced by biliverdin in combination with two other pigments never before found in bird eggs: red-orange tripyrrolic uroerythrin and yellow-brown tetrapyrrolic bilirubin, respectively.

(a)

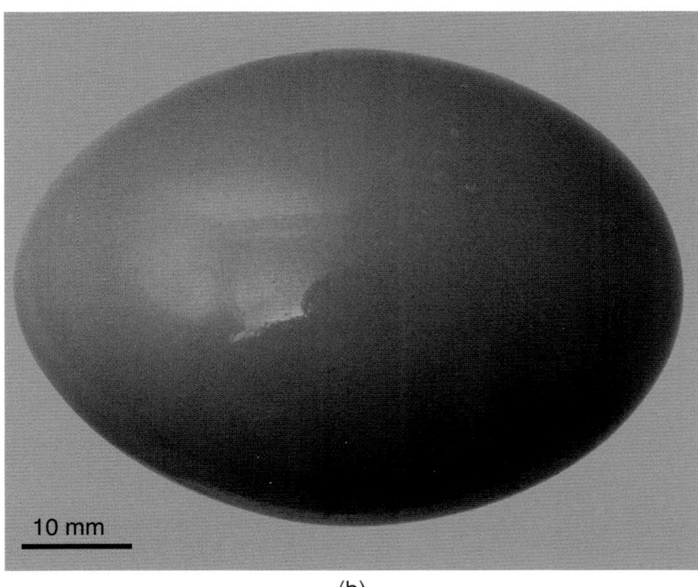

(b)

FIGURE 5.6

Eggs from (a) the spotted nothura and (b) the elegant-crested tinamou. [Museum of Comparative Zoology, Harvard University, specimens 355101, 367141, © President and Fellows of Harvard College.]

FIGURE 5.7

The largest bird eggs were those of the extinct elephant bird. The smallest are those of hummingbirds. [Museum of Comparative Zoology, Harvard University, specimens 352251, 352276, 367177, 359951, © President and Fellows of Harvard College.]

SIZE AND SHAPE OF EGGS

Smaller birds lay smaller eggs and larger birds lay larger eggs. But just how small and how large?

The smallest living bird, the bee hummingbird, which lives in Cuba, is so small, just two and a half inches long, that it is often mistaken for a bee. Its eggs are roughly the size of a coffee bean (¼ inch or 6 mm). That's dramatically smaller than a large chicken egg (2¼ inches or 55 mm). The largest living bird, the ostrich, lays eggs that are about the size of a cantaloupe (7 inches or 175 mm), and the now-extinct ten-foot-tall elephant bird of Madagascar had much larger eggs still—a bit larger than a football (13.5 inches or 340 mm) (figure 5.7).

The shape of bird eggs ranges from the nearly spherical eggs of owls to the highly asymmetric eggs of murres and sandpipers, which are much more pointed at one end than the other (figure 5.8). Just how varied are the shapes of eggs? A remarkable survey, by Cassie Stoddard, at Princeton University, and her colleagues, of over 49,000 eggs from about 1,400 bird species, about 13% of all species in the world, in the collections of the Museum of Vertebrate Zoology at the University of California, Berkeley, characterized egg shape by how elliptical an egg is (roughly, how elongated) and how asymmetric it is (roughly, how different the radii are at the blunt and pointed ends) (figure 5.9). Using a digital image of each egg that the museum already had in its database, the researchers fit a curve to the outer profile of each egg and then used the equation of that curve to determine the egg's ellipticity and asymmetry.

The full range of egg shapes can be seen on the plot of egg ellipticity against egg asymmetry that Stoddard and her colleagues produced; the plot also includes photographs of examples of extreme egg shapes and more typical egg shapes superimposed on the plot (figure 5.10). The nearly spherical egg of the eastern screech-owl is shown in the lower left-hand corner of the plot (low ellipticity and low asymmetry). The maleo egg at the top left corner of the plot is elliptical, with both ends roughly equally blunt (low asymmetry). The least sandpiper egg at the extreme right side of the plot has the largest difference in the radii at the blunt and pointed ends (the highest asymmetry). Many species of birds have eggs with a shape similar to that of the graceful prinia.

An egg attains its final shape within the oviduct, after the double membrane is added over the albumen but before the calcium carbonate shell is deposited onto the membrane. The shape of the egg is determined by the shape of the membrane

FIGURE 5.8

Egg shape ranges from roughly spherical (eastern screech-owl) to elliptical (maleo) to the asymmetric (common murre). [Museum of Comparative Zoology, Harvard University, specimens 352837, 357806, 352291, 354365, 359951, 361788, 353571, © President and Fellows of Harvard College.]

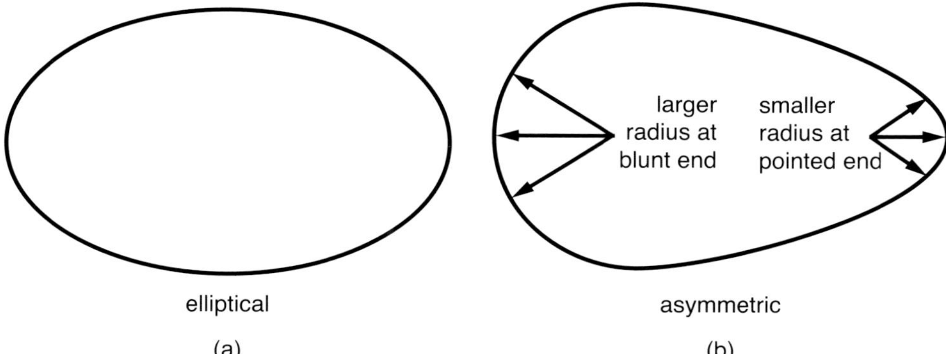

FIGURE 5.9

For each egg they examined, Stoddard and her colleagues quantified how (a) elliptical and (b) asymmetric the egg was.

within the oviduct, which is added just before the calcium carbonate shell is deposited.

Having mapped out the range of egg shapes, Stoddard and her colleagues then asked how this range is produced within the oviduct. Their hypothesis turns out to be fascinating: muscular tension within the oviduct applies pressures to the egg membrane, and the shape of the membrane (and final egg) is determined by both the pressures acting over it and the properties of the membrane itself (for instance, its thickness and its stiffness, which can vary along the length and around the circumference of the egg, depending on the type, arrangement, and amount of collagen fibers within it).

Using biomechanical modeling, which allowed them to vary the membrane properties and the pressure acting over the membrane, Stoddard and her colleagues were able to generate the entire range of observed shapes of eggs (figure 5.11). It remains to be seen whether the thickness and stiffness of egg membranes in eggs of different shapes also vary in the way that the model suggests.

Stoddard and her colleagues were able to show how eggs are shaped in different ways. But the question remains: Why do eggs have such a large range of shapes? Various explanations have been proposed.

Spherical eggs have advantages. A sphere has the smallest surface area per unit volume of any shape, which means that for a constant volume of egg, less calcium carbonate is required for a spherical shape than for any other. The small surface area per unit volume also means that a spherical egg loses less heat (when the adult is not sitting on it) than a nonspherical egg of the same volume. And spheres are the most structurally efficient shell shape.

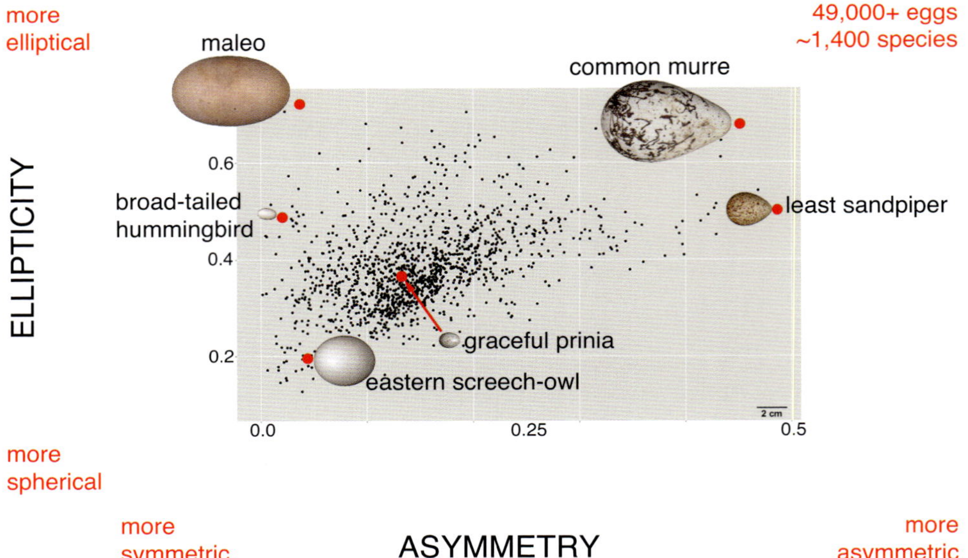

FIGURE 5.10

The shapes of eggs from 1,400 species of birds, with some representative eggs highlighted. [Adapted from Stoddard et al. (2017), with permission; egg photos from the Museum of Comparative Zoology, Harvard University, specimens 352837, 352291, 354365, 359951, 361788, 353571, © President and Fellows of Harvard College.]

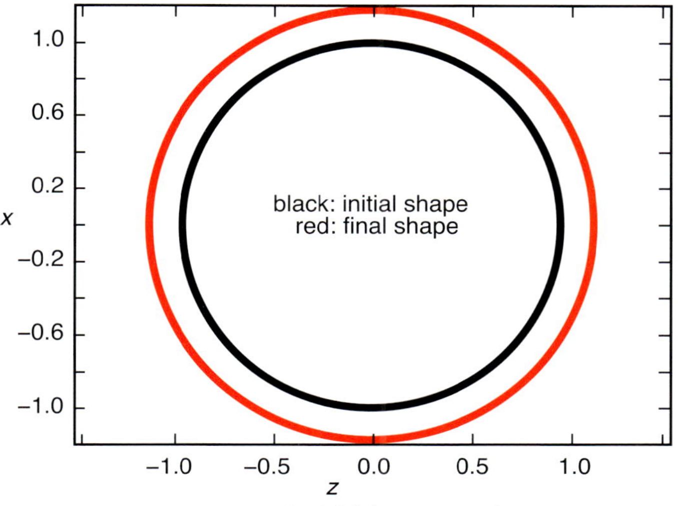

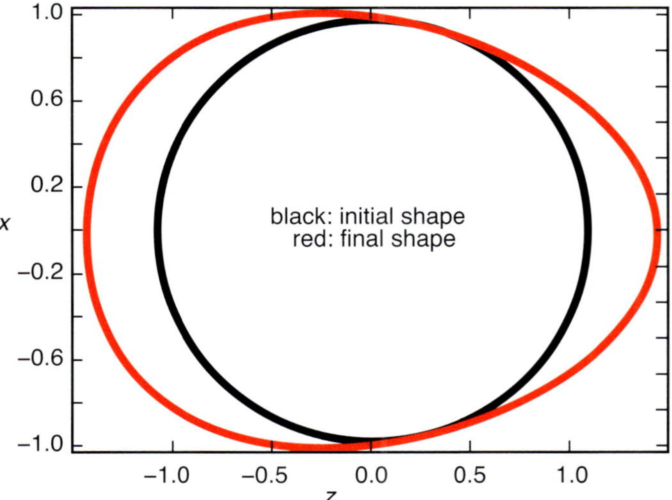

FIGURE 5.11

Computer modeling is able to generate the full range of egg shapes observed in bird eggs, starting with an initially spherical egg membrane. (a) A model of a membrane with the same thickness and stiffness everywhere and uniform pressure within the membrane gives a spherical egg. (b) A model with a thinner membrane on the right side of the egg leads to an asymmetric egg. [After Stoddard et al. (2017), fig. S9.]

If that's the case, then why aren't all eggs spherical?

One thought is that nonspherical eggs may assist in heat transfer from the adult to the egg because they have a larger surface area than spherical eggs of the same volume. Shorebirds with asymmetric eggs, like the least sandpiper and the spotted sandpiper, often lay four eggs that they arrange with the pointed ends toward the center of the nest, packing the eggs closer together than would be possible with spherical eggs, again increasing heat transfer from adult to egg (figure 5.12).

It was long thought that the very asymmetric eggs of guillemots, which have one end much pointier than the other (similar to the common murre egg on the right of figure 5.8), meant that if the eggs rolled, they would do so in a tight circle—an important characteristic that reduced the chances of eggs rolling off the cliff ledges where guillemots nest. But Tim Birkhead, of the University of Sheffield, who has studied guillemots and their eggs for over 40 years, has debunked this idea in his book *The Most Perfect Thing: Inside (and Outside) a Bird's Egg*. Birkhead notes that the studies that support the tight-circle claim either used emptied-out museum eggshells or model eggs made of plaster, neither of which accurately replicated the behavior of actual guillemot eggs, which in fact *don't* roll in tight circles.

Birkhead suggests two other explanations for the shape of the guillemot egg.

First, he writes, it's possible that the shape causes more of the shell to be in contact with the hard ground on which it rests (compared with a spherical shell of the same volume), reducing the internal stresses within the egg and making it less likely to be damaged as the adult sits on the egg or as the egg rolls over rough ground.

The second explanation relates to guillemots' nesting behavior: the birds nest in colonies on cliff ledges, packed closely together and even touching (often at 20 pairs of birds per square meter, and occasionally as high as 70 pairs per square meter; a square meter is about 20% larger than a square yard). As a result, there is often fecal matter on guillemot eggs, especially at the pointed end, which is often in contact with the ground. The shape of a guillemot egg reduces the amount of fecal matter on the blunt end of the egg, which has the highest shell porosity, allowing gas exchange as the chick develops.

The large database of egg shape produced by Stoddard and her colleagues allowed a more comprehensive approach to answering the question of what accounts for egg shape. For each bird species in their egg shape database, they gathered data on biometric, life history, and environmental parameters: their data set included adult body mass, diet, clutch size, nest type, nest location, chick developmental mode (*precocial*, leaving the nest within a day or two of hatching, versus

spotted sandpiper nest and eggs

FIGURE 5.12

Shorebirds often arrange their eggs with the pointy ends toward the center of the clutch to increase the surface area in contact with the incubating adult, increasing the heat transferred to the eggs. [Alamy.]

altricial, depending on their parents for food and protection), and environmental details (latitude, temperature, average precipitation). They also included a measure of flight ability in their data. Bird species with stronger and more frequent flight tend to have narrower and more pointed wingtips (think swallows, swifts, and terns). Ornithologists characterize this by the "hand-wing index"—that is, the ratio of the length to the width of the wingtip. Stronger fliers tend to have a higher hand-wing index.

When Stoddard and her colleagues measured the hand-wing index on museum specimens of each species in their database of egg shapes, they found that egg shape was associated with egg length and the hand-wing index: birds that lay longer eggs and those with a high hand-wing index tend to lay eggs that are more asymmetric or elliptical. Egg shape was not typically associated with the other factors they looked at.

Strong fliers need to have aerodynamically streamlined bodies with a narrow pelvis to reduce drag during flight. The narrow pelvis likely requires a narrow egg: for two species of birds of the same size, with eggs of the same volume, the species that is the stronger flyer may have a narrower pelvis, requiring a narrower and longer egg that is more elliptical or asymmetric. Swallows, swifts, and terns, all strong fliers, have elongated, asymmetric eggs. Egg shape appears to be driven, at least in part, by adaptations for flight.

This is one of those results that isn't at all obvious—but once it's explained, it makes so much sense that other researchers say to themselves enviously, "Why didn't I think of that?" The idea that egg shape might be related to flight ability was actually suggested in 1991 by John Iverson and Michael Ewert, but nobody gave it much thought until Stoddard and her colleagues picked it up again and did the hard work of quantifying the shapes of tens of thousands of eggs and relating their shapes to physiological and environmental factors. Studies like theirs would be impossible without the extensive collections of natural history museums. We'll see another example of the usefulness of such collections at the end of this chapter.

STRENGTH OF EGGSHELLS

Eggshells have to be strong enough to resist the weight of the adult bird incubating the egg without fracturing, yet weak enough for the chick to break out when ready to hatch. As the chick skeleton develops within the embryo, it absorbs calcium from the shell, reducing its thickness, making it easier for the chick to hatch out.

We've all cracked chicken eggs open in the kitchen and think of eggshells as fragile, but they're surprisingly strong when the load is distributed over the shell. It's actually possible for a barefoot person to walk (carefully) on chicken eggs still in their carton without breaking them.[2]

Just how strong are eggs? One way to measure an egg's strength is to compress it at the top and bottom of the egg, distributing the load at the ends of the egg with layers of foam with egg-shaped cutouts for the ends of the egg to fit into (figure 5.13). Chicken eggs loaded in this way fail at a surprisingly high load of about 160 pounds (70 kg).

When you compress a hollow sphere at the top and bottom, you produce compressive forces in the sphere. What's not so obvious is that you also produce tensile forces: as you push down on the sphere, the middle tends to bulge out ever so slightly, which puts it in tension (figure 5.14).

Brittle materials that fail suddenly, like eggshells, are weaker in tension than in compression (because tension pulls small cracks in the material apart, leading to fracture, while compression simply pushes the faces of the crack together). Thus an egg loaded in compression, as in figure 5.13, fails in tension.

The tensile strength of an eggshell depends on the load applied at the top of the egg as well as the cross-sectional area of the eggshell, which in turn depends on the diameter and thickness of the eggshell. The measured strength of the eggshells in the experiments in figure 5.13 is about 2,200–4,400 pounds per square inch (15–30 MPa in the metric system). For comparison, this is about one-fifth the tensile strength of solid bone.

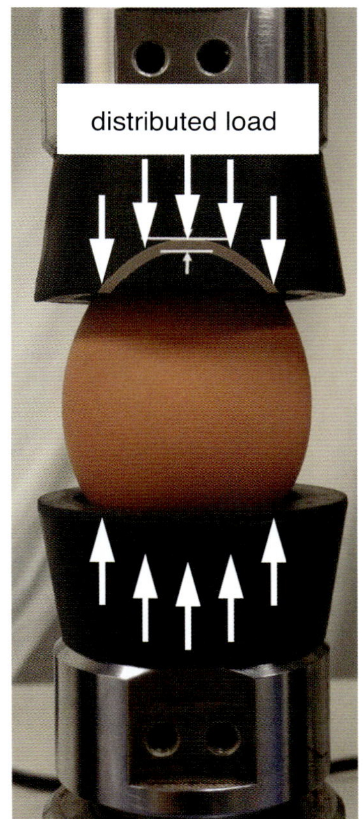

FIGURE 5.13

Testing the strength of a chicken egg. Chicken eggs break at about 160 pounds (70 kg) when loaded uniformly across the shell. [From Hahn et al. (2017), reproduced with permission.]

FIGURE 5.14

A force, F, applied to the top of a spherical shell produces internal compression and tension forces within the shell.

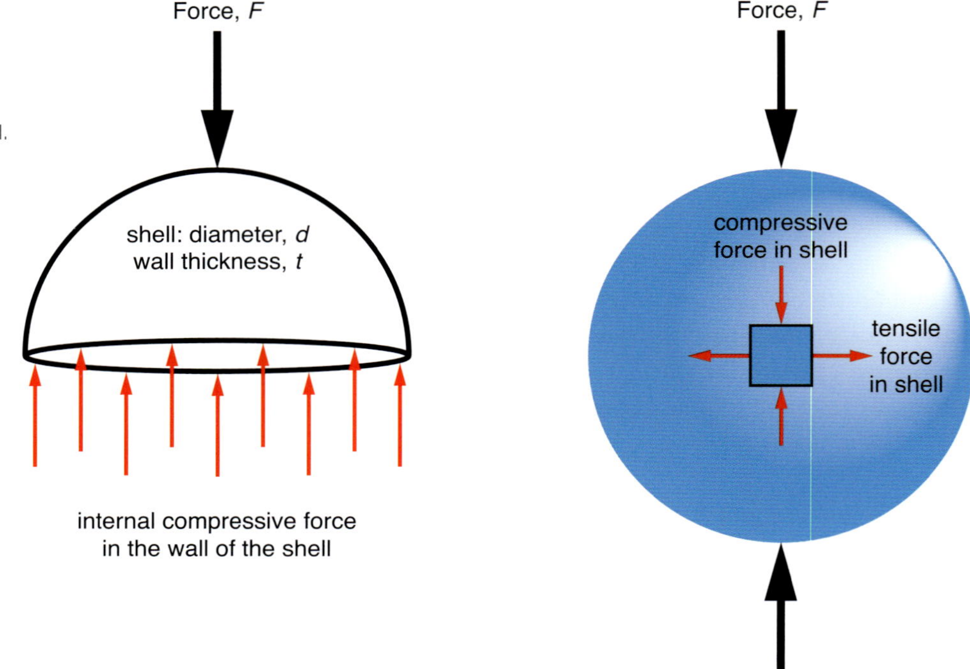

An abnormal reduction in eggshell thickness can have disastrous consequences. After World War II, DDT (dichlorodiphenyltrichloroethane), which had been developed during the war, was sprayed intensively on agricultural crops. By the late 1950s, it was apparent that the spraying was killing large numbers of birds from pesticide poisoning, especially in species at the top of the food chain, such as peregrine falcons and ospreys, which were eating prey with higher concentrations of the toxins. Even the iconic bald eagle was suffering a drastic reduction in numbers. Birds were also failing to reproduce successfully due to abnormally thin eggshells caused by DDT; in some cases, the eggshells were so thin that they fractured under the weight of the incubating adult bird.

Using Eurasian sparrowhawk eggs in natural history museums that were collected between 1870 and 1980, Ian Newton and Margaret Haas, of the Monks Wood Experimental Station, in England, showed that the thickness of the eggshells decreased dramatically in the late 1940s, just as DDT was introduced for agricultural use (figure 5.15).

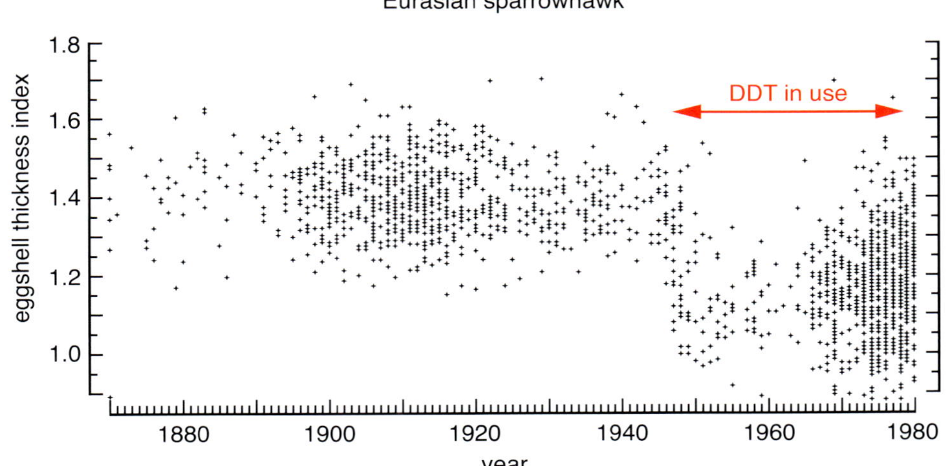

FIGURE 5.15

The introduction of DDT after World War II dramatically reduced the thickness of eggshells of Eurasian sparrowhawks. [After Newton and Haas (1984).]

FIGURE 5.16

Eggshells of Eurasian sparrowhawks and peregrine falcons were weaker, failing at lower loads, after the introduction of DDT. [After Cooke (1979).]

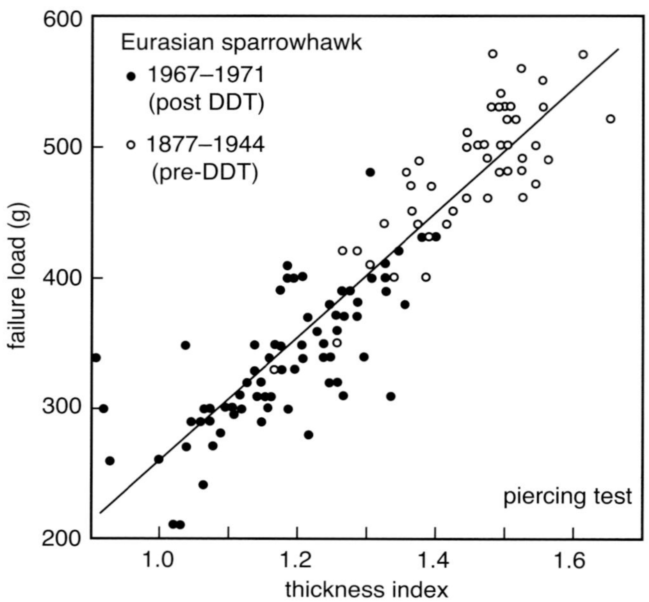

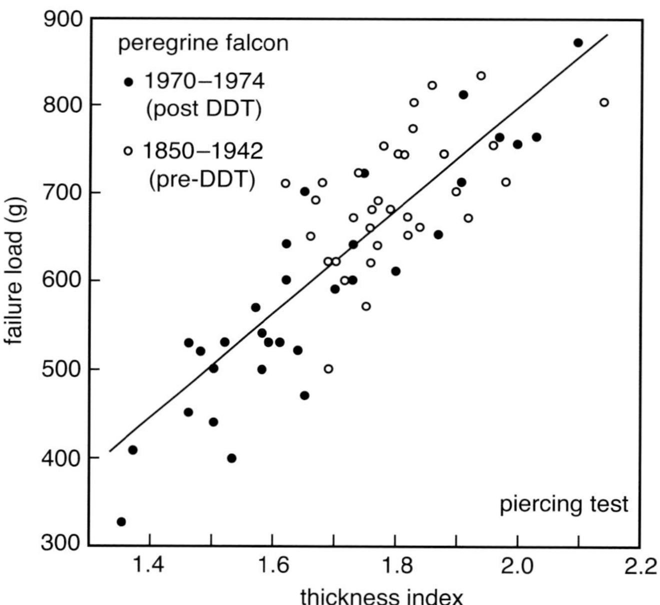

In another study, Arnold Cooke, also of the Monks Wood Experimental Station, compared the thickness and piercing strength of Eurasian sparrowhawk and peregrine falcon eggshells collected between 1850 and 1944, before DDT was in widespread agricultural use, with those of eggshells collected between 1967 and 1974, when it was widely used. (Again, the eggs used in the study came from natural history museum collections.) The piercing tests were done using a machine like that in figure 5.13. The researchers simply placed the egg on a flat bottom plate and attached a rod with a sharp point to the top plate, which was then moved downward until the sharp point pierced the eggshell. The newer set of eggshells, collected between 1967 and 1974, were thinner and broke at a lower load than the older set, collected before DDT was introduced (figure 5.16).

Rachel Carson's book *Silent Spring*, published in 1962, alerted the world to the dangers of DDT; 10 years later, DDT was banned in the United States. Since then, the numbers of peregrine falcons, osprey, and bald eagles have recovered, in large part because the thickness of their eggs has returned to that prior to the use of DDT. I've seen peregrine falcons and bald eagles at Jamaica Pond, right in the city of Boston, something that would not have been possible 50 years ago.

As you may have noticed, I love going to Harvard's Museum of Comparative Zoology to look at their ornithological collections. Seeing eggs of familiar species, I find myself instantly transported back to seeing the birds in the field, delighted to see what their eggs look like, something I don't get to experience when I'm out birding. Tiny white tree swallow eggs call to mind tree swallows darting over Jamaica Pond in late March, one of the first birds to return from their wintering grounds. Spherical white screech-owl eggs: a pair of screech-owls, one gray, one reddish, that one year nested in a hole in a rotted tree branch at the end of my street. Bigger snowy owl eggs: snowies standing at the top of the sand dunes at Plum Island, north of Boston, on the lookout for prey. A southern cassowary egg, huge and buff-colored, dappled in the palest green: walking with a guide though the Daintree rainforest in Queensland, Australia, seeing a southern cassowary just in front of us, almost too close for comfort. Hummingbird eggs, at the other extreme, tiny, white, looking delicate in their nests: ruby-throated hummingbirds probing honeysuckle flowers in my garden. Eggs of so many different colors, sizes, and shapes, all so rich in the way they function, and all so wonderful.

REFERENCES

Birkhead T (2012). *Bird Sense: What It's Like to Be a Bird*. Bloomsbury.

Birkhead T (2016). *The Most Perfect Thing: Inside (and Outside) a Bird's Egg*. Bloomsbury.

Birkhead TR, Thompson JE, and Biggins JD (2017). Egg shape in the Common Guillemot *Uria aalge* and Brunnich's Guillemot *U. lomvia*: not a rolling matter? *J. Ornith.* **158**, 679–685.

Birkhead TR, Thompson JE, Jackson D, and Biggins JD (2017). The point of a Guillemot's egg. *Ibis* **159**, 255–265.

Carson R (1962). *Silent Spring*. Houghton Mifflin.

Cassey P, Maurer G, Lovell GP, and Hanley D (2011). Conspicuous eggs and colourful hypotheses: testing the role of multiple influence on avian eggshell appearance. *Avian Biology Research* **4**, 185–195.

Cooke AS (1979). Changes in egg shell characteristics of the Sparrowhawk (*Accipiter nisus*) and Peregrine (*Falco peregrinus*) associated with exposure to environmental pollutants during recent decades. *J. Zoology London* **187**, 245–263.

Ehrlich PR, Dobkin DS, and Wheye D (1988). *The Birder's Companion: A Field Guide to the Natural History of North American Birds*. Simon and Schuster/Fireside.

Gill FB (2007). *Ornithology*. Third edition. WH Freeman.

Hahn EN, Sherman VR, Pissarenko A, Rohrbach SD, Fernandes DJ, and Meyers MA (2017). Nature's technical ceramic: the avian eggshell. *J. Roy. Soc. Interface* **14**, 20160804.

Hamchand R, Hanley D, Prum RO, and Bruckner C (2020). Expanding the eggshell colour gamut: uroerythrin and bilirubin from tinamou (*Tinamidae*) eggshells. *Scientific Reports* **10**, 11264.

Hauber ME (2014). *The Book of Eggs: A Life-Size Guide to the Eggs of Six Hundred of the World's Bird Species*. University of Chicago Press.

Higginson TW (1862). The life of birds. *Atlantic Monthly* **10** (September), 368–376.

Iverson JB and Ewert MA (1991). Physical characteristics of reptilian eggs and a comparison with avian eggs. In *Egg Incubation: Its Effect on Embryonic Development in Birds and Reptiles*, edited by Deeming DC and Ferguson MWJ, 87–100. Cambridge University Press.

Lovell PG, Ruxton GD, Langridge KV, and Spencer KA (2013). Egg-laying substrate selection for optimal camouflage by quail. *Current Biology* **23**: 260–264.

Lovette IJ and Fitzpatrick JW, eds. (2016). *The Cornell Lab of Ornithology Handbook of Bird Biology*. Third edition. Wiley.

Newton I and Haas MB (1984). The return of the Sparrowhawk. *British Birds* **77**, 47–70.

Purcell R, Hall LS, and Corado R (2008). *Egg and Nest*. Belknap Press.

Ratcliffe DA (1967). Decrease in eggshell weight in certain birds of prey. *Nature* **215**, 208–210.

Stoddard MC, Yong EH, Akkaynak D, Sheard C, Tobias JA, and Mahadevan L (2017). Avian egg shape: form, function and evolution. *Science* **356**, 1249–1254.

6

FLIGHT: WEIGHT, UPWARD FORCE, AND LIFT

Learning the secret of flight from a bird
was a good deal like learning the secret of magic from a magician.
Orville Wright

People have always been entranced by bird flight. Since at least the time of Leonardo da Vinci, there were attempts to design a machine that would allow people to fly. When the Wright brothers finally succeeded, with the Wright Flyer, it was in part because of the lessons they had learned during years of studying bird flight. They started by reading everything they could. Books such as J. Bell Pettigrew's *Animal Locomotion; or Walking, Swimming and Flying, with a Dissertation on Aeronautics* (1873), and Otto Lilienthal's *The Problem of Flying* (1894), emphasized the need to observe birds in flight to understand how their wings worked. They learned that the upward force keeping a bird aloft against gravity arises from the flow of air over the curved shape of its wings. And they learned that the faster air flows over the wing, the more upward force it produces.

The Wright brothers also learned from the work of Louis Pierre Mouillard, a Frenchman who lived much of his life in Algeria and Egypt. In *Empire of the Air: An Ornithological Essay on the Flight of Birds*, first published in 1881, Mouillard wrote with wonder about vultures in Africa soaring great distances and suggested that they might be a model for the design of gliding aircraft. The brothers took this suggestion to heart. From their home in Dayton, Ohio, they rode their bicycles to the Pinnacles, a deep river gorge with unusual pyramid-like rock formations, to watch turkey vultures soar on the currents of air rising up out of the gorge.[1] They saw that raising the tip of one wing while lowering the other caused a bird to turn and realized that a plane with a flexible wing, one that could be twisted so that one wingtip rose while the other dipped, would be easier to steer than one with more rigid wings. They understood that it wasn't enough to simply stay aloft; a pilot needed to be able to control the altitude of the plane and steer it.

When the Wright brothers wanted to start experimenting with flight, they sought out a location that had steady, strong winds (to fly into, to get more upward force) and sandy beaches (in case they crashed). From wind velocity data they received from the US Weather Bureau, they selected Kitty Hawk, on the Outer Banks of North Carolina. And on each trip to Kitty Hawk to experiment with their plane, they watched the birds: gulls, gannets, turkey vultures, hawks, and more. They observed how they moved their wings to understand why they moved them in the way that they did. As one of the locals, John T. Daniels, told *Collier's Weekly* later on, "They'd stand on the beach for hours at a time just looking at the gulls flying, soaring, dipping. They seemed to be interested mostly in gannets. They would watch the gannets and imitate the movements of their wings with their arms and hands. They could imitate every movement of the wings of those gannets; we thought they were crazy." (The article referred to the Wright brothers as the "Dayton Gannets.")

The Wright brothers' observations of bird flight were key in helping them learn how to design and fly a plane. They began by building a glider with a rudder and flexible wings that could be twisted to raise one side and lower the other, and they flew it like a kite, controlling it from the ground with cables. Once they understood how to control it, they next flew the glider, with one of the brothers on board, getting it aloft by running along a downward sandy slope and into the wind until it took off, which allowed them to develop their steering skills. Once they felt confident that they could control the glider, they added an engine. Their first powered flight, with Orville Wright at the controls, was made on the morning of December 17, 1903. The Wright brothers learned well from the birds.

Bird flight is a wonder (figure 6.1). Mallard ducks and Canada geese can fly 40 mph (65 km/hr) in still air; with a strong tailwind they can reach 60 mph (nearly 100 km/hr). Raptors and vultures soar upward thousands of feet on columns of warm

FIGURE 6.1

Birds in flight: (a) mallard, (b) Canada geese in V formation, (c) broad-winged hawk soaring, (d) peregrine falcon stoop amid a flock of European starlings, (e) ruby-throated hummingbird hovering, (f) common murre swimming, and (g) a murmuration of European starlings. [All Alamy except (d) Rob Palmer (www.falconphotos.com).]

FIGURE 6.2

The forces acting on a bird in horizontal flight: upward force, weight, thrust, and drag. [After Alexander (2002).]

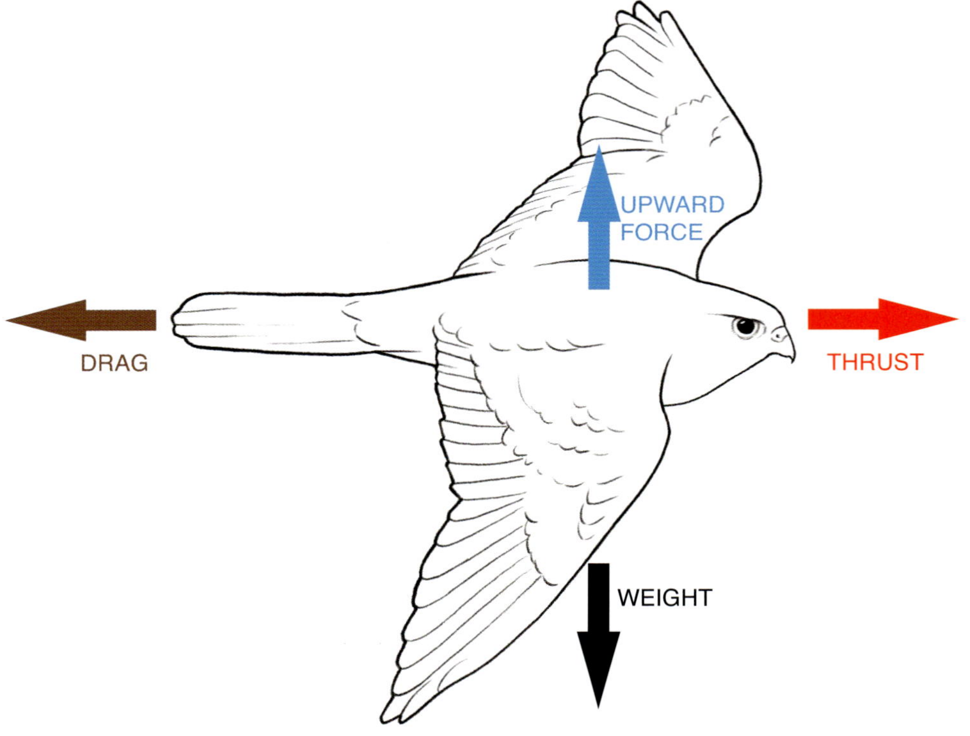

air, then glide forward for miles effortlessly. Peregrine falcons dive at up to 200 mph (320 km/hr) to maim and kill smaller birds in flight, then maneuver to grasp their prey in their talons, often before it hits the ground. Hummingbirds hover while collecting nectar at flowers, flapping their wings roughly 45 times a second, and can even fly backward. Penguins and alcids, such as guillemots and murres, flap their wings to "fly" underwater. And in the fall and winter European starlings form huge flocks—of tens or even hundreds of thousands—that seem to dance across the sky at dusk. How do they do it all?

To answer that question, we'll need to get a little technical. In this chapter and the next, we'll look at how birds fly by examining the forces involved in flight (figure 6.2). In this chapter, we'll look at *weight*, which is the force of gravity acting on the bird, and the *upward force* balancing weight that results from differences in air pressure on the top and bottom surfaces of the wing.

WEIGHT

One winter after I'd moved to Boston, I returned home to Canada and went for a walk with my mother through the woodland trails at the Royal Botanical Gardens in Burlington, Ontario. When we reached a wooden bridge over a small creek, we stopped to listen to chickadees calling from their perches on the nearby bushes. My mother knew the spot and had come prepared. She put a few sunflower seeds on her palm and held it out, and within a few moments a chickadee landed on it, took one, and flew off (figure 6.3). My mother was tickled pink, as she liked to say, delighted at how light and delicate the bird was. She handed me some seeds, and soon a chickadee landed on my hand, its tiny feet and claws tickling my fingertips. I knew birds were light, but until that moment I hadn't quite realized how light. A typical chickadee weighs less than half an ounce or 14 grams—one eighth of a stick of butter.

Birds typically weigh one-third to one-half the weight of a mammal of comparable size (figure 6.4). For instance, a house sparrow and an eastern chipmunk are both about 6 inches (150 mm) long; the sparrow's weight is roughly a third that of the chipmunk. A rock pigeon and an eastern gray squirrel are both about 11 inches (0.28 m) long; the pigeon's weight is about half that of the squirrel. An American crow and an eastern cottontail rabbit are both about 17 inches long (0.43 m); the crow's weight is about a third that of the rabbit. (The bird lengths include the tail while the mammal lengths do not.)

FIGURE 6.3

Mum feeding a chickadee. [WL Gibson.]

FIGURE 6.4

Comparison of weights of birds and mammals: (a) house sparrow, (b) eastern chipmunk, (c) pigeon, (d) eastern gray squirrel, (e) American crow, and (f) eastern cottontail rabbit. Data from *The Sibley Field Guide to Birds* and the *Peterson Field Guide to Mammals of America North of Mexico*; bird length includes tail; mammal length does not. [All Alamy.]

(a) house sparrow
L = 6.25 in W = 1 oz

(b) eastern chipmunk
L = 5–6 in W = 2.3–4.5 oz

(c) pigeon
L = 12.5 in W = 0.56 lb

(d) eastern gray squirrel
L = 8–10 in W = 0.75–1.6 lb

(e) American crow
L = 17.5 in W = 1 lb

(f) eastern cottontail rabbit
L = 14–17 in W = 2–4 lb

So how do birds achieve their low weight?

Chipmunks, squirrels, and rabbits all have relatively thin fur coats, and it's not difficult to get a sense of how big the body of each one is beneath the fur. It's harder with birds, because feathers can account for up to half the volume of a bird. We see the outer contour feathers of a bird and imagine that its body is just beneath. But birds have a thick layer of down between their bodies and their outer feathers. And, as we saw in chapter 2, a bird's down is only a few percent solid keratin—the rest is entrapped air. In short, birds' bodies beneath their feathers aren't as big as they appear to be.

There are even more ways that feathers reduce weight. As a bird flies, turning this way and that, its flight feathers bend and twist in response to variations in air pressure over the vanes of the feathers (figure 6.5). The internal structure of the shaft of the flight feathers, a roughly square tube filled with a foam (figure 6.6), allows them to be lightweight while still providing the needed resistance to bending and twisting.

We saw in chapter 3, on bones, that the internal tensile and compressive forces in a bent beam are largest at the top and bottom surfaces of the beam, and that cross sections that have most of the material at the top and bottom of the beam, such as a sandwich beam or an I-beam, reduce the weight of the beam for a given deflection or failure load (compared with a solid square cross section). When a square tube bends, it acts like an I-beam, with the top and bottom edges acting as the flanges and the vertical edges acting as the web.

How much weight savings does the foam-filled square tubular cross section provide? Compared with a solid square cross section of the same bending stiffness (one that deflects the same amount under a given load) or of the same bending strength (one that fails at the same load), the flight feather cross section shown in figure 6.6 weighs about half as much. Here I've assumed that the cross section shown is approximately a square tube with 2 mm edges that are 0.2 mm thick, filled with a foam that is about 5% solid—but it actually has thicker top and bottom sides and thinner vertical sides, which saves even more weight. (Note that the cross section tapers along the length of the feather shaft and at the very tip is elliptical rather than square in cross section, so that this is just a ballpark estimate of weight saving.)

Similarly, the internal deformations and forces in a twisted shaft are largest at the outer perimeter of the shaft (figure 6.7). The most efficient, lowest-weight shaft for resisting twisting (torsion) is a thin-walled circular tube. The feather shaft's square tube is a compromise that allows it to effectively resist both bending and twisting.

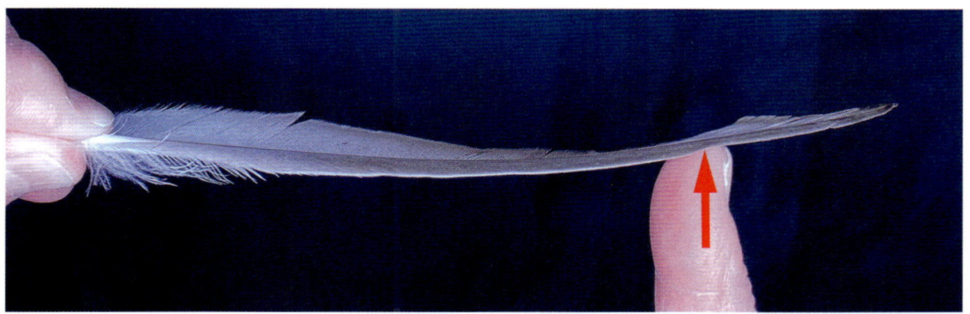
(a) bending

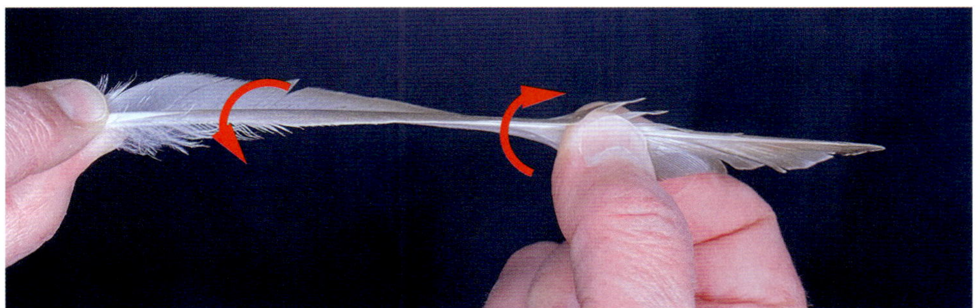
(b) twisting

FIGURE 6.5

Loads on flight feathers: (a) bending and (b) twisting.

FIGURE 6.6

(a) Mallard flight feather and (b) cross section of the feather shaft, showing the outer tube of solid keratin filled with foamy keratin. [Photos: Dr. Isaac Cabrera.]

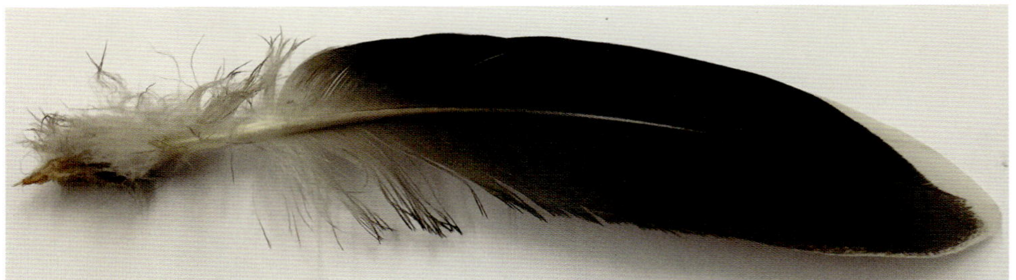

(a)

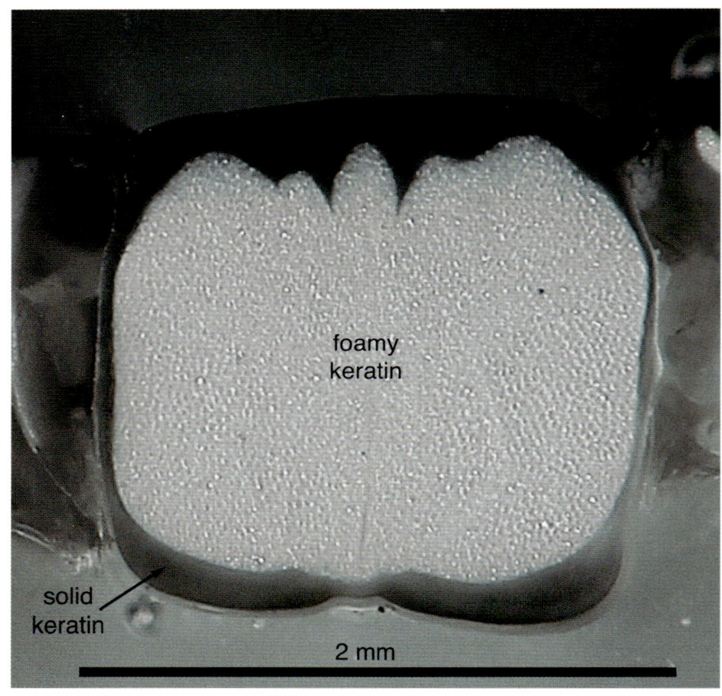

(b)

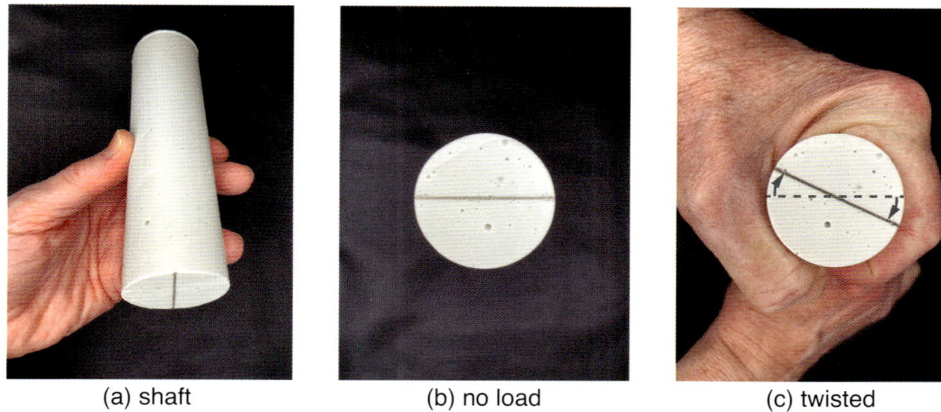

FIGURE 6.7

(a, b) A rubber shaft with a circular cross section. (c) When the shaft is twisted, the line rotates from the dotted line to the solid line. The deformation is largest at the outside of the shaft and zero at the center.

FIGURE 6.8

Kinking in a drinking straw.

And what about the foamy keratin core within the feather shaft? What role does it play? If you take a drinking straw and bend or twist it, it kinks (figure 6.8). Kinking is common in hollow tubes, especially ones with thin walls, like the feather shaft (or the drinking straw). Filling the tube with foam increases its resistance to kinking by supporting the tube wall, pushing back against kinking.

Foam-filled tubes are common in nature. The stems of grasses have an outer, nearly solid tube supported by an inner layer of foam-like cells (in grasses, the interior foam-like layer typically does not completely fill the tube) (figure 6.9a). Studies of the mechanics of plant stems reveal that the foam-like layer increases the resistance of the stem to kinking, compared with a tube of the same weight and diameter without the foam-like layer.

Porcupine quills are modified hairs that have a similar structure to that of feather shafts: a thin, outer tube of solid keratin filled with foamy keratin (figure 6.9b). Porcupine quills and feather shafts both need to prevent kinking. And they must also reduce heat loss along the quill or shaft. It is likely that the interior core of foamy keratin in feather shafts and porcupine quills both resists kinking and reduces heat loss, helping to keep the bird or porcupine warm.

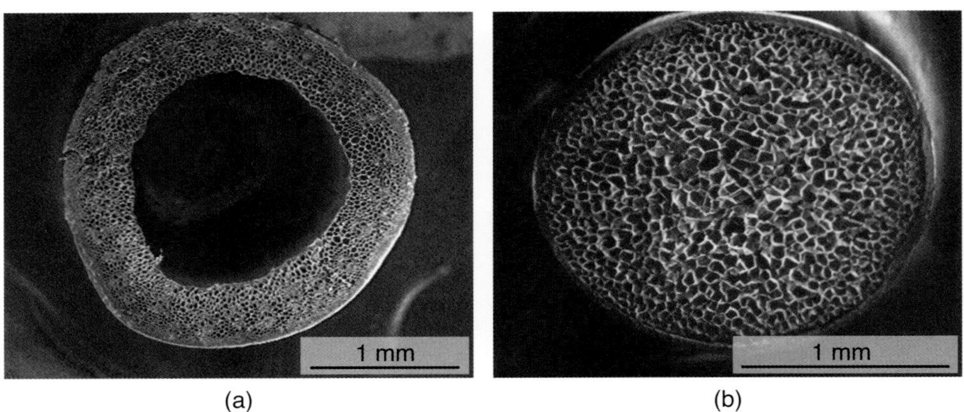

FIGURE 6.9

Scanning electron microscope images of (a) a grass stem and (b) a porcupine quill. Both are circular tubes with a foam-like core. [From Karam and Gibson (1995), © Elsevier, with permission.]

FIGURE 6.10

Space shuttle fuel tank, showing the interior latticework to prevent kinking of the outer shell. The tanks were 30 feet in diameter and 150 feet long. [Photo: NASA.]

When engineers design large cylindrical tubular structures, such as the legs of offshore oil platforms or the space shuttle fuel tanks, they include an inner gridwork of plates to avoid kinking; the gridwork acts in a way similar to that of the foamy core in plant stems, porcupine quills, and feather shafts (figure 6.10).

As we saw in chapter 3, the skeleton of a modern bird is adapted for reduced weight in several ways. The tail is made up of lightweight feathers anchored in a short, stubby tailbone. The bill lacks teeth and is largely made up of a spongy, foam-like bone surrounded by thin layers of solid bone and keratin. Groups of neighboring bones are fused together (as is the case with "hand" bones, groups of vertebrae, and the pelvic bones), allowing them to be lighter than unfused bones for the same stiffness and strength. Flat plate-like bones (like the sternum) and curved, shell-like bones (like the skull and fused pelvic bone) have a sandwich structure—with thin, outer skins of solid bone separated by a core of spongy bone—that acts like an I beam, leading to significant weight reductions. (The approximate calculations in chapter 3 suggest weight reductions of over 50% compared with a hypothetical solid bone of the same stiffness and strength.) The long bones of the wing and leg are

FIGURE 6.11

Birds' sex organs shrink out of breeding season. (a) Testis mass and (b) ovarian follicle diameter. [After Ball and Ketterson (2008).]

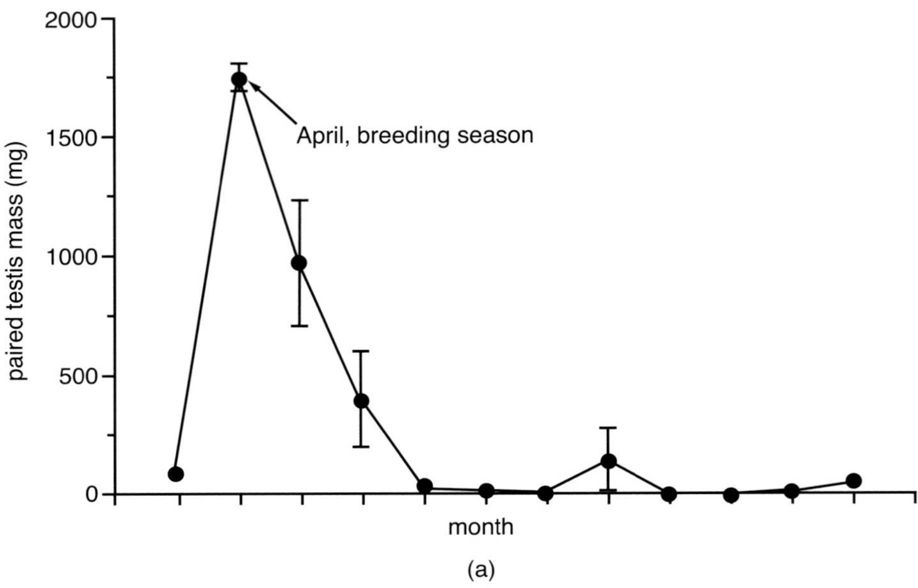

(a)

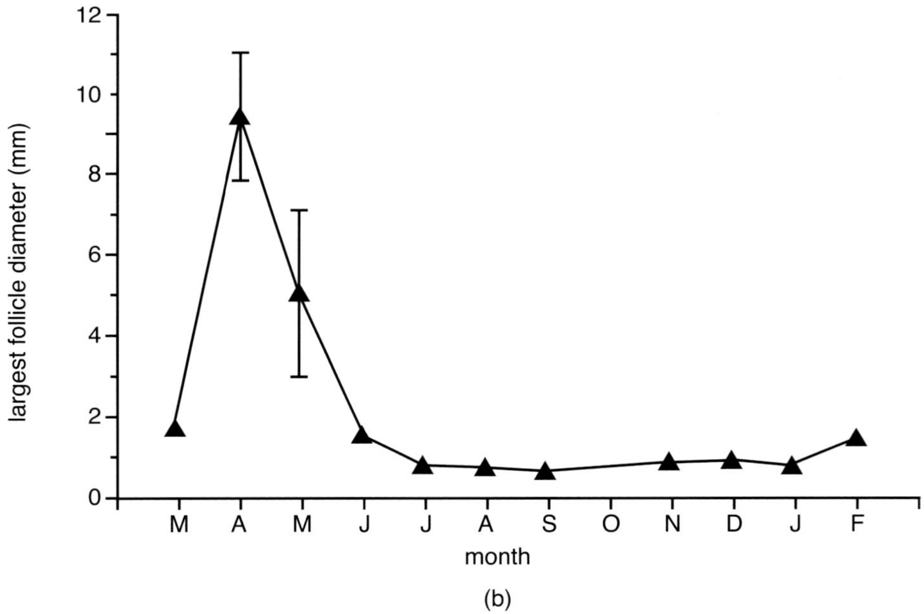

(b)

tubular, with thinner walls relative to their diameter than those of mammal bones; the approximate calculations in chapter 3 suggest that for the same stiffness and strength this reduces weight by roughly 30%. And some of the long bones are filled with air (functioning as reservoirs in the bird's respiratory system), rather than marrow, further reducing their weight.

Birds have even more adaptations to reduce their weight. Females lay eggs rather than giving birth to more developed, and heavier, young. The eggs go through the oviduct one at a time, so that the female doesn't have to fly with multiple fully developed eggs. Most female birds have only one oviduct rather than two, as in mammals. In birds that breed seasonally, the sex organs shrink, sometimes dramatically, once the breeding season is over; the testes in male starlings, for example, decrease in volume by a factor of over 1,000 (figure 6.11)! The ovaries in female birds, too, shrink after breeding. Unlike mammals, birds do not have a urinary bladder; instead, they excrete a slurry of uric acid and feces through a single opening, the cloaca, which often falls on vehicles and other surfaces as a white paste.

All of these adaptations decrease a bird's weight and, in turn, reduce the amount of upward force needed to maintain a bird aloft during flight.

UPWARD FORCE

Probably the most puzzling thing about flight to birders is how a bird stays aloft. Where does the upward force balancing the weight of the bird come from (figure 6.12)? To answer these questions, let's first look at a simpler situation than the gliding or flapping flight of a bird: air flowing over a stationary engineering wing in a wind tunnel. Gliding and flapping flight will be discussed in the next chapter.

The path of the air around the wing is shown in the top photograph of figure 6.13. Smoke particles, added into the airflow upstream of the wing through a vertical pipe with evenly spaced holes, make the curved paths of the air visible; these curved paths are called "streamlines." The lower drawing shows streamlines schematically.

Some distance ahead of the airfoil, the air moves at a constant velocity and the streamlines are parallel and unaffected by the wing. As the air approaches and reaches the wing, it is forced to curve around it, inducing changes in the velocity and pressure of the air throughout the flow around the wing. The changes in velocity and pressure associated with moving the air in these curved paths are what generate the upward force.

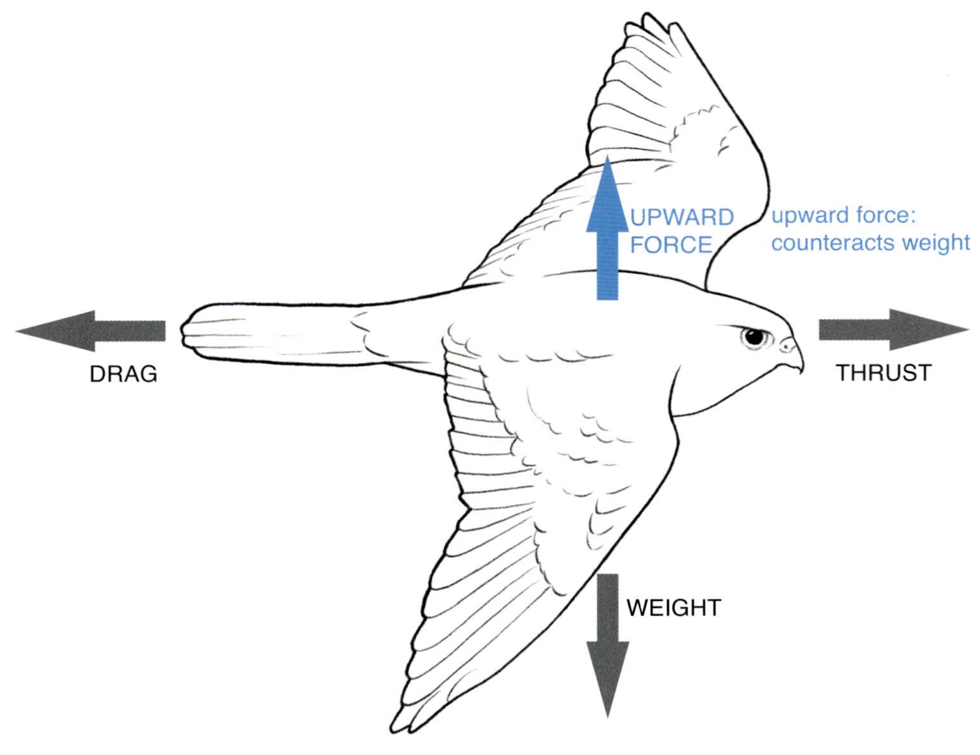

FIGURE 6.12

Forces acting on a bird in horizontal flight: upward force. [After Alexander (2002).]

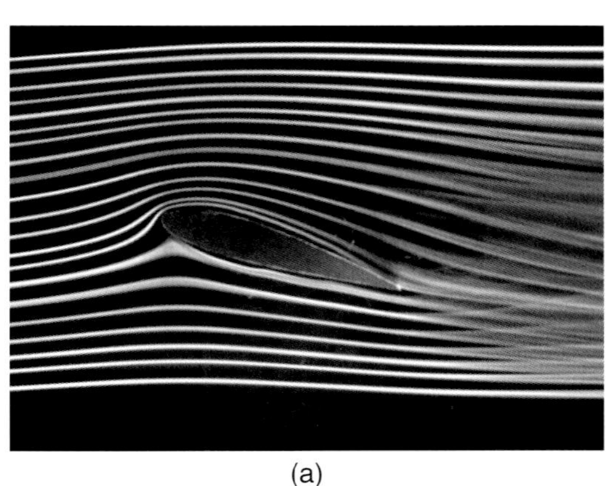

(a)

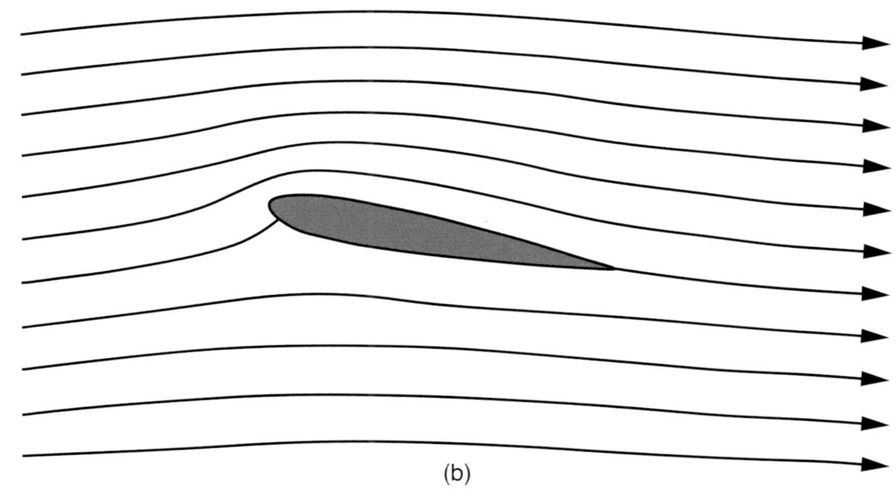

(b)

FIGURE 6.13

Airflow over a wing showing streamlines. (a) Wing in a wind tunnel; (b) schematic. [(a) Babinsky (2003). © IOP Publishing. Reproduced with permission. All rights reserved. (b) Diagram after https://en.wikipedia.org/wiki/Airfoil, public domain.]

Since it's hard to imagine forces acting on air moving in a curved path, let's start with a more familiar, analogous situation: the force acting on a ball on a string, swung in a circle at a constant velocity (figures 6.14 and 6.15). The string pulls on your hand (this is what you feel as the force), the string itself is in tension, and the force on the ball acts inward, pulling the ball toward the center of the circle at each instant, preventing it from flying off at a tangent to the circle.

Applying this idea to air flowing in a circular path, we can replace the ball with a small cube of air, of mass *m* (figure 6.16). In this case, there is no string as there is with the ball. For the air to move in a circular path, there still has to be an inward force acting on it, forcing the air along the circular path. (Without an inward force, the air would simply move in a straight line.)

The inward force arises from pressure differences acting on the outer and inner surface of the cube of air (figure 6.17). For the force on the cube of air to act inward (like the force on the ball on the string), the pressure on the outer surface of the cube has to be higher than that on its inner surface. As air moves in a circular or curved path, the pressure is always *lower on the inner side of the curve* than on the outer side.

Think of a tornado: objects are sucked toward the center of a tornado because the air pressure in the center of the cone is lower than that at the outer wall of the tornado (figure 6.18). In the lower schematic, the air on the outside of each circle is at a higher pressure than the air on the inside of each circle.

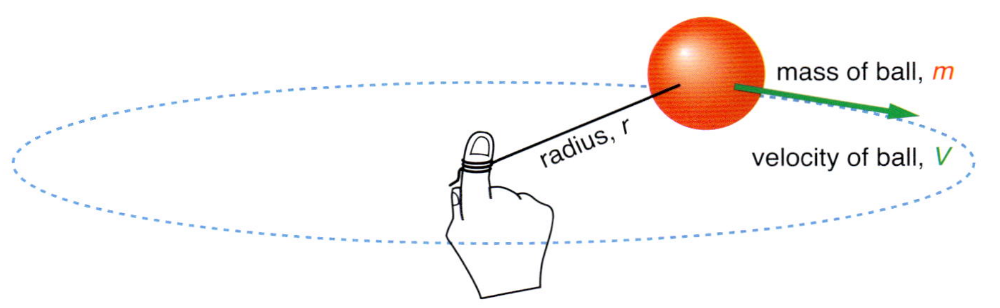

FIGURE 6.14

A ball on a string moving in a circular path. The force on the ball increases with increasing mass, *m*, of the ball and velocity, *V*, of the ball. The force decreases as the radius, *r*, of the circle increases.

force:
increases with mass of ball, *m*
increases with velocity of ball, *V*
decreases with radius of circle, *r*

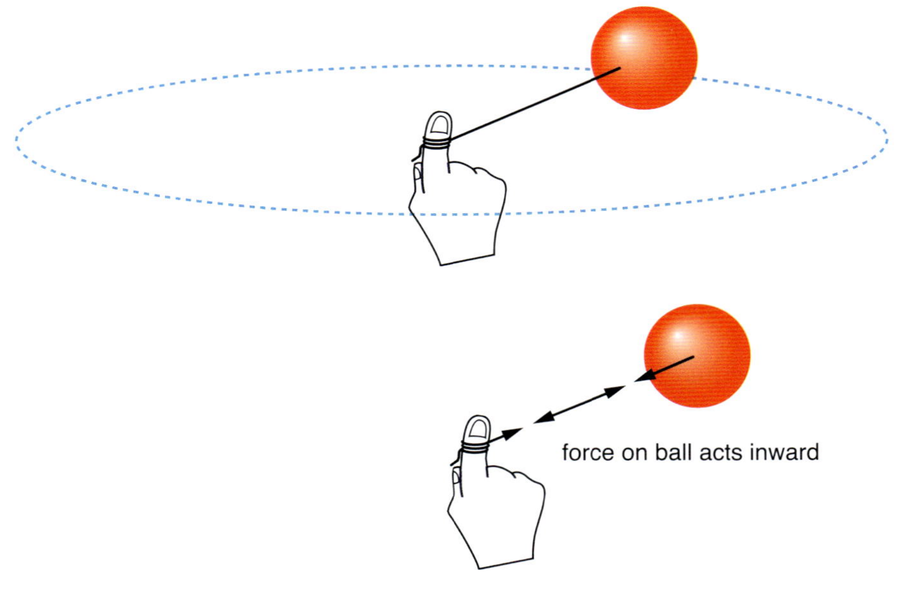

FIGURE 6.15

The string pulls on the hand. The string is in tension. The force on the ball acts inward.

FIGURE 6.16

For a cube of air moving in a circular path, there is also an inward force on it, acting toward the center of the arc. If there was no force acting, the air would move in a straight, horizontal path.

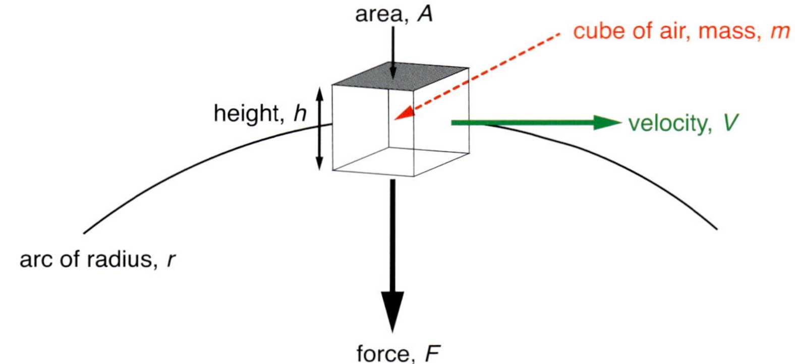

FIGURE 6.17

The inward force arises from pressure differences on the outer and inner surface of the cube of air.

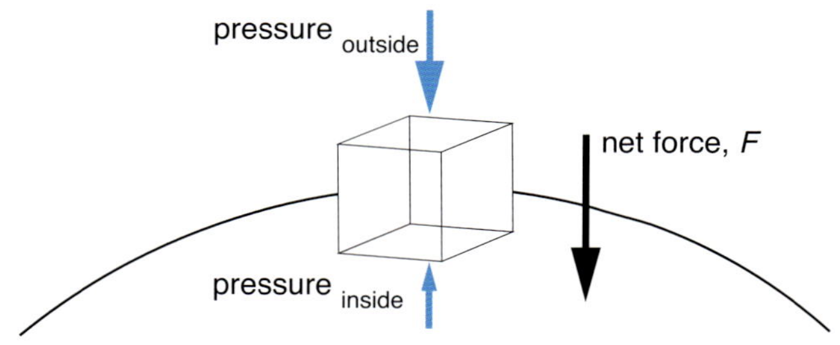

FIGURE 6.18

Tornadoes have lower pressure inside the cone than at their outer wall. This is what sucks objects into the center of a tornado. [Photo: Alamy.]

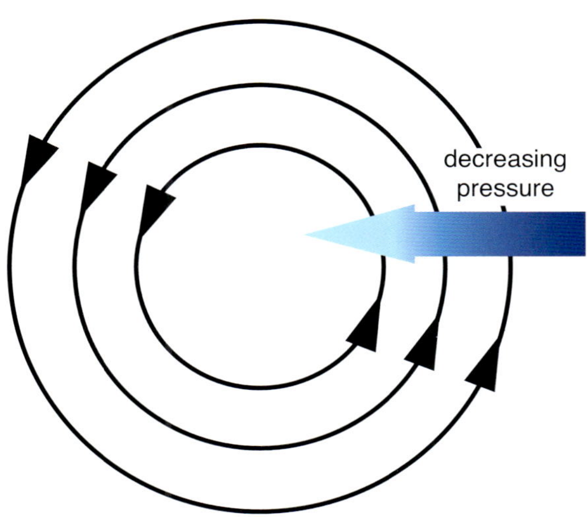

Now let's think in more detail about air flowing over a curved wing (figure 6.19). Far away from the wing, either above or below it, the air is at atmospheric pressure (p_{atm}) (typically, a little less than 15 pounds per square inch [psi] or about 1 bar or 0.1 MPa in the metric system). First, let's consider the air above the wing (figure 6.19). Remembering that the pressure is *lower* on the *inside* of the curved path of airflow, we can see that the pressure at a point A, just above the wing, is less than the atmospheric pressure high above the wing ($p_A < p_{atm}$).

Next, let's look at the air below the wing (figure 6.20): point B, just below the wing, is on the outside of the curve of the streamlines, indicating that the pressure at B is greater than the atmospheric pressure well below the wing ($p_B > p_{atm}$). The net result is that the pressure above the wing, at A, is lower than that below the wing, at B: this pressure difference gives an *upward force*. The pressure differences and resulting upward force counteract the weight of the bird to keep it aloft.

It has been reported that peregrine falcons sometimes exploit the upward force in an unexpected way. They're well known for killing other birds in flight with their incredibly fast, steep dives known as "stoops," in which they reach speeds of up to 200 mph (about 320 km/hr). Usually, they dive down with their wings tucked in to reduce drag. But sometimes they flip upside down, rotating about their length, and extend their wings (figure 6.21). With their extended wings upside down, the upward force is directed downward, increasing their downward acceleration and speed. They then reverse themselves to be right side up to hit and kill their prey with their feet.

While the variations in pressure and air velocity over the entire region around the wing are complex and require computational methods to determine exactly, we can get a glimpse of how air pressure and velocity are related quantitatively by considering the differences in pressure (a) across curved streamlines (box 6.1) and (b) along a curved streamline (box 6.2). The explanation of the upward force given above and in the two boxes is based on an article by Holger Babinsky of Cambridge University. Further details on computational methods for determining the pressures and velocities around a wing can be found in the articles on aerodynamic lift by Doug McLean, a former aeronautical engineer at Boeing. The references are at the end of this chapter.

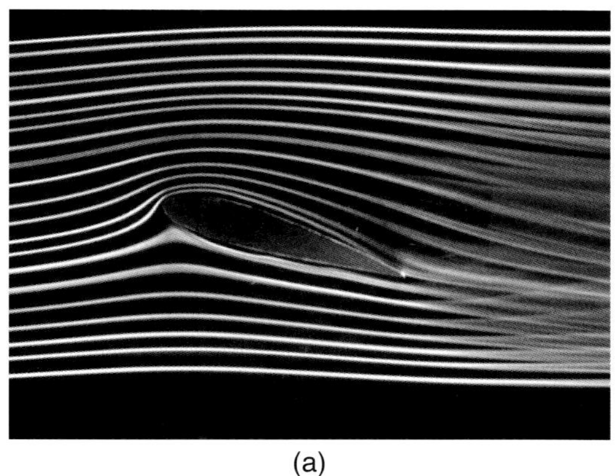
(a)

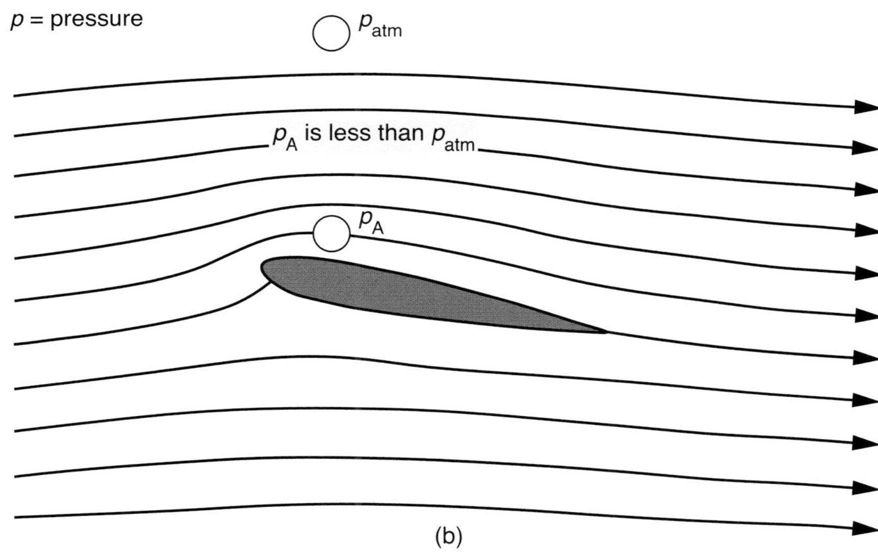

(b)

FIGURE 6.19

(a) Streamlines around a wing in a wind tunnel. (b) The air pressure far above the wing is atmospheric pressure, p_{atm}. The point A, just above the wing, is on the inside of the curve of the streamlines and the pressure at A, p_A, is lower than atmospheric pressure, p_{atm}. [(a) Babinsky (2003). © IOP Publishing. Reproduced with permission. All rights reserved. (b) Diagram after https://en.wikipedia.org/wiki/Airfoil, public domain.]

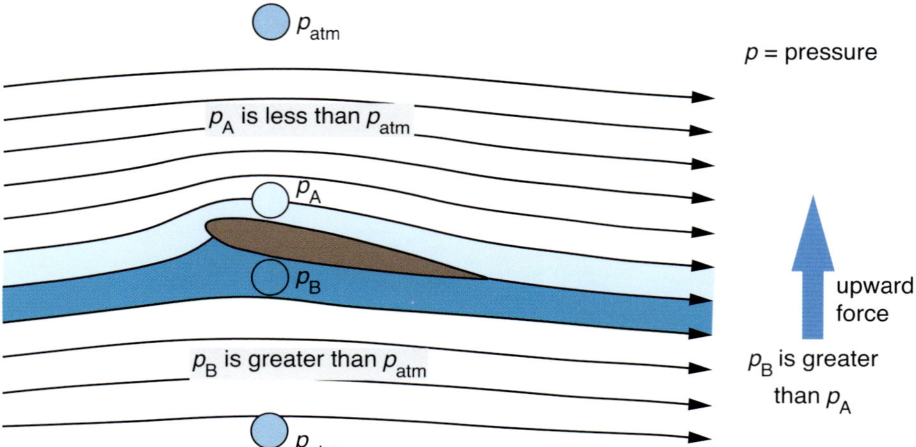

FIGURE 6.20

The air pressure far below the wing is atmospheric pressure, p_{atm}. The point B, just below the wing, is on the outside of the curve of the streamlines and the air pressure at B is higher than atmospheric pressure, p_{atm}. The net effect is that the pressure below the wing (at B) is higher than the pressure above the wing (at A), producing an upward force. Darker blue is higher pressure; lighter blue is lower pressure. [Diagram after https://en.wikipedia.org/wiki/Airfoil, public domain.]

FIGURE 6.21

A peregrine falcon flying upside down. With the wings upside down, the normal "upward force" acts downward, increasing the speed of the falcon. [Photo by Rob Palmer (www.falconphotos.com).]

BOX 6.1: PRESSURE DIFFERENCES ACROSS CURVED STREAMLINES

Let's return to the force acting on a ball on a string, swung in a circle at a constant velocity (figure 6.14). If you've ever swung a ball on a string, you'll probably have a physical sense that the force in the string increases as the mass of the ball increases and as the velocity of the ball increases. And you'll have the sense that as the length of the string (or radius of the circle) increases, the force in the string decreases.

The equation relating the force on the ball to its mass, m_{ball}, and velocity, V_{ball}, and to the length of the string, or radius of the circle, r_{circle}, can be found by thinking about the units of each parameter: the units of mass, velocity, and radius have to combine to be consistent with the units of force. The metric system of units is simpler for this purpose, so we'll use it.

Mass has units of kilograms (kg), velocity has units of meters/second (m/s), and radius has units of meters (m). Force is equal to mass times acceleration, and acceleration is the change in velocity with time. For instance, accelerating from 0 to 10 m/s in 5 seconds is an acceleration of 10/5 = 2 m/s^2. So the units of force are mass × acceleration (kg m/s^2).

Combining so that the units are the same on both sides of the equation gives

$$Force = \frac{mass_{ball} \cdot Velocity_{ball}^2}{radius_{circle}}$$

$$F = \frac{m_{ball} \cdot V_{ball}^2}{r_{circle}}$$

The string pulls on your hand (this is what you feel as the force), the string itself is in tension, and the force on the ball acts inward, pulling the ball toward the center of the circle at each instant, preventing it from flying off at a tangent to the circle (figure 6.15).

Applying this idea to air flowing in a circular path, we can replace the ball with a small cube of air of mass m (figure 6.16). In this case, there is no string as there is with the ball. For the air to move in a circular path, there still has to be an inward force acting on it, forcing the air along the circular path. (Without an inward force, the air would simply move in a straight line.)

The inward force arises from pressure differences acting on the outer and inner surface of the cube of air (figure 6.17). For the force on the cube of air to act inward (like the force on the ball on the string), the pressure on the outer surface of the cube has to be higher than that on the inner surface. As air moves in a circular path, the pressure is always *lower on the inner side of the curve* than on the outer side.

The force acting on the cube of air of mass m, moving with velocity V along a circular path of radius r, is:

$$F = \frac{mV^2}{r}$$

The mass of air in the cube is its density, ρ (ρ is the Greek letter rho), times the volume of the cube ($A \times h$):

$$F = \frac{\rho AhV^2}{r}$$

Pressure is force divided by area (in pounds per square inch or psi in the US system, or newtons per square meter in the metric system). The pressure difference across the cube of air is then:

$$(pressure_{outside} - pressure_{inside}) = \rho V^2 \left(\frac{h}{r}\right)$$

This calculation is for air moving at a constant velocity in a circular path, but the same ideas apply for air moving along any curved path. The pressure difference *across* streamlines depends on the density of air, the square of the velocity of the air, and a geometrical factor that incorporates the radius of curvature of the curve.

BOX 6.2: PRESSURE DIFFERENCES FROM AIR MOVING ALONG A STREAMLINE (BERNOULLI'S EQUATION)

Bernoulli's equation, formulated by Daniel Bernoulli, an eighteenth-century scientist, gives the relationship between pressure p and air velocity V, at two points 1 and 2 *along a streamline*, for air of a density ρ:

$$p_1 + \frac{\rho V_1^2}{2} = p_2 + \frac{\rho V_2^2}{2}$$

The difference in pressure between the two points depends on the air density and the square of the velocity of the air at each point. This is a simplified form of Bernoulli's equation, derived using various assumptions (for instance, that there is little difference in the height of the two points, that the flow does not change with time, and others); for air flowing over the wing of a bird, all of the assumptions are valid.

We've seen how smoke particles added into the airflow upstream of a wing in a wind tunnel illustrate the streamlines (figure 6.13). If the smoke particles are pulsed into the airflow, it's possible to see that the air speeds up over the top of the wing and slows down as it passes across the bottom of the wing. Higher speed also corresponds to more closely spaced streamlines, while slower speed corresponds to more widely separated streamlines (figures 6.13, 6.22).

If we follow a single streamline above the wing (figure 6.22), the air velocity at point 2 is higher than at point 1, where the streamlines are parallel and unaffected by the wing. Bernoulli's equation then tells us that the air pressure at point 2 is less than that at point 1. Similarly, if we follow a single streamline beneath the wing, where the air velocity slows down, we can see that the pressure at point 4 is greater than at point 3. Since the pressure far in front of the wing is constant ($p_1 = p_3$), the pressure below the wing, p_4, is greater than that above the wing, p_2, producing an upward force.

Writing Bernoulli's equation for the two streamlines:

$$p_1 + \frac{\rho V_1^2}{2} = p_2 + \frac{\rho V_2^2}{2}$$

$$p_3 + \frac{\rho V_3^2}{2} = p_4 + \frac{\rho V_4^2}{2}$$

And since, far ahead of the wing, the pressure and velocity are constant ($p_1 = p_3$ and $V_1 = V_3$), we have:

$$p_2 + \frac{\rho V_2^2}{2} = p_4 + \frac{\rho V_4^2}{2}$$

The difference in pressure between points 2 and 4, above and below the wing, depends on the density of air and the squares of the velocities of the air at the two points.

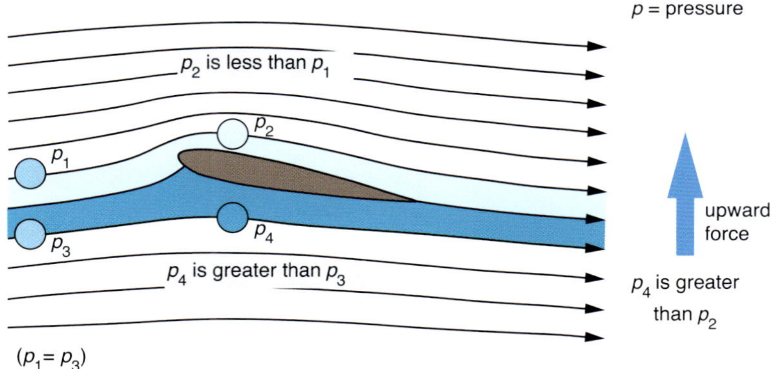

FIGURE 6.22

Pressure difference above and below a wing from Bernoulli's equation. [Diagram after https://en.wikipedia.org/wiki/Airfoil, public domain.]

LIFT

Lift is related to the upward force, but the two are not identical. The lift force is defined to act perpendicular to the direction of the airflow over a wing; as a hawk glides gradually downward or a Canada goose flaps its wings, the airflow over their wings is not horizontal and lift does not act vertically. Based on what we've learned so far, we'll first calculate the magnitude of the lift force. In the next chapter, we'll see how lift and drag interact in the gliding and flapping flight of birds.

We've seen how airflow around a wing produces changes in the air velocity and pressure which lead to an upward force counteracting the weight of a bird. In the boxes, we've shown that the difference in pressure above and below the wing depends on the air density, ρ, and the square of the air velocity, V, as ρV^2.

The magnitude of the lift force depends on the difference in pressures above and below the wing times the area of the wing (since pressure is force/area). Lift also depends on the geometry of the wing: the shape of the wing and the angle of the wing relative to the direction of the airflow, referred to as the angle of attack (figure 6.23). A lift coefficient, C_L, accounting for the shape of the wing and the angle of attack, is measured in wind tunnel experiments. Combining everything, the lift force is then:

$$F_{lift} = C_L \frac{\rho_{air} A_{wing} V_{air}^2}{2}$$

(The factor 2 is introduced to relate this to the kinetic energy/volume of air [density × velocity^2/2], also called the dynamic pressure.)

The dependence of lift on the density of air can have unexpected consequences. In the summer of 2017, during a heatwave in Arizona in which temperatures reached 119°F (48°C), the *Washington Post* ran a headline that read, "It's so hot in Phoenix that airplanes can't fly."[2] The density of air decreases with increasing temperature (which is why hot air rises). During that summer, the temperatures in Phoenix were so high, and the density of the air consequently so low, that at the speeds the planes could attain before reaching the end of the runways, the airflow over their wings could not produce sufficient lift to take off. Fortunately, the birds in Phoenix don't need a runway and didn't have that problem.

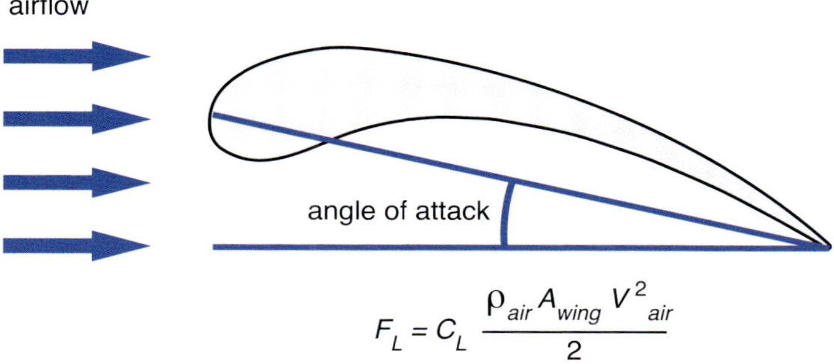

FIGURE 6.23

The lift force depends on the density of air, ρ_{air}, the area of the wing (as seen from above), A_{wing}, the velocity of air, V_{air}, and the lift coefficient, C_L. The lift coefficient depends on the shape of the wing and the angle of attack that the wing makes with the direction of airflow. [Diagram after https://en.wikipedia.org/wiki/Airfoil, public domain.]

I'm at Wellfleet, Massachusetts on Cape Cod, high up on the dunes overlooking the Atlantic on a sunny summer morning, the clear sky an intense blue, watching gulls fly along the shore, their wings held out in two curves, just the way you see in simplistic sketches of gulls. It's exhilarating just being here, looking out over the ocean, seeing the waves come ashore.

As the gulls fly by, I think of the wonder of flight, for so long a mystery. The remarkable fact that what keeps a bird aloft is simply the air flowing over the curved shape of its wings, inducing lower pressure above the wings and higher pressure below the wings, generating enough upward force to balance the bird's weight. And I think of the features of birds that make them lightweight, reducing the amount of upward force needed: the airy down feathers; the efficient internal structures of their bones; their toothless bills; their ability to produce and lay one egg at a time; and the reduction of some of their internal organs.

REFERENCES

Alexander DE (2002). *Nature's Flyers: Birds, Insects and the Biomechanics of Flight*. Johns Hopkins University Press.

Babinsky H (2003). How do wings work? *Physics Education* **38**, 497–503.

Ball GF and Ketterson ED (2008). Sex differences in the response to environmental cues regulating seasonal reproduction in birds. *Phil. Trans. Roy. Soc.* **B363**, 231–246.

Bissonette TH and Chapnick MH (1930). Studies on the sexual cycle in birds. II The normal progressive changes in the testis from November to May in the European starling (*Sturnus vulgaris*), an introduced, non-migratory bird. *American Journal of Anatomy* **45**, 307–343.

Burt WH and Grossenheider RP (1976). *A Field Guide to Mammals of America North of Mexico*. Houghton Mifflin.

Flindt R (2006). *Amazing Numbers in Biology*. Translated by Neil Solomon. Springer.

Gill FB (2007). *Ornithology*. Third edition. WH Freeman.

Jakab P (2013). Leonardo da Vinci and flight. On the National Air and Space Museum website (accessed January 1, 2024), https://airandspace.si.edu/stories/editorial/leonardo-da-vinci-and-flight

Karam GN and Gibson LJ (1995). Elastic buckling of cylindrical shells with elastic cores—I Analysis. *Int. J. Solids and Structures* **32**, 1259–1283.

Lilienthal O (1894). *The Problem of Flying*. US Government Printing Office.

Lovette IJ and Fitzpatrick JW, eds. (2016) *The Cornell Lab of Ornithology Handbook of Bird Biology*. Third edition. Wiley.

McCullough D (2015). *The Wright Brothers*. Simon and Schuster.

McLean D (2018a). Aerodynamic lift, part 1: the science. *Physics Teacher* **56**, 516–520.

McLean D (2018b). Aerodynamic lift, part 2: a comprehensive physical explanation. *Physics Teacher* **56**, 521–524.

Mouillard LP (1881). *L'empire de l'air: essai d'ornithologie appliquée à l'aviation*. G Masson. Translated as "Empire of the Air: An Ornithological Essay on the Flight of Birds," in Smithsonian Institution, Annual Report, 1892, 397–463.

Pettigrew J B (1873). *Animal Locomotion; or Walking, Swimming and Flying, with a Dissertation on Aeronautics*. Henry S King.

Roberson JA and Crowe CT (1975). *Engineering Fluid Mechanics*. Houghton Mifflin.

Saunders WO (1927). Then we quit laughing. *Colliers*, September 17.

Sibley D (2003). *The Sibley Field Guide to Birds of Eastern North America*. Knopf.

Tennekes H (2009). *The Simple Science of Flight: From Insects to Jumbo Jets*. Second edition. MIT Press.

Wright O. Quoted from Malsbury E (2020). How we lifted flight from bird evolution. *Smithsonian Magazine*, December 17.

7

FLIGHT: DRAG AND THRUST

Birds are the most perfectly trained gymnasts in the world and are specially well fitted for their work.
Wilbur Wright

Like the last chapter, this one is a bit more technical than most of the book. The reward is that you'll learn some of the secrets of bird flight. With an understanding of lift and drag, the resistance created as air moves past the bird, we'll see why geese fly in V formations and how soaring birds such as broad-winged hawks and albatrosses exploit air currents over land and ocean to travel thousands of miles. We'll learn how flapping flight generates forward thrust. And in a final section on specialized types of bird flight, we'll look at how hummingbirds modify their flapping to enable them to hover and even fly backward; how penguins and alcids such as murres and guillemots "fly" underwater with their wings; and how scientists track and study the mesmerizing murmurations of starlings.

DRAG

Drag is just a resistance to forward motion (figure 7.1). We've all experienced drag in the form of wind resistance. Riding my bicycle into a headwind, I feel the wind pushing against me: that's drag. Somewhat annoyingly, the harder and faster I pedal, the higher the net speed of the wind against me, and the higher the drag. If you hold your hand (carefully) outside the window of a moving car, you can feel the drag force against your hand. There is more resistance when your hand is held with the palm vertical than with the palm horizontal; drag also depends on the area that is perpendicular to the moving air (called the projected area). Drag acts parallel to the direction of the airflow.

In general, as air moves past a stationary object, the streamlines move smoothly over the top and bottom surfaces, but there is turbulence in the wake behind the object (figure 7.2). The pressure ahead of the object is higher than that in the turbulent region; this pressure difference gives rise to *pressure drag*. The larger the turbulent wake, the higher the drag; the cross section of the cylinder shown in the figure has a larger turbulent wake and higher drag than the cross section of the airfoil. For a wing moving through air, pressure drag is one of three main sources of drag.

Another major source of drag is from the friction of the air moving past the surface of the bird. Right at the surface, on the molecular level, the air is still. Over a thin layer above the surface, the velocity of the air increases until it reaches the velocity of the general airflow. As the air gains velocity over that thin layer, it has to overcome viscous resistance, leading to friction drag. *Friction drag, also called skin friction drag*, depends on the viscosity of air, the velocity of the bird through the air, and the surface roughness of the bird.

The final source of drag, *induced drag*, is from vortices that form at the wingtips. At the tip of a wing, the higher pressure beneath the wing meets the lower pressure above the wing, causing air to circulate around the tip, forming vortices (figure 7.3). Just beyond of the tip of the wing, the air moves upward and then circulates back downward closer to the wingtip. The downward motion on the wingtip of the bird pushes down against the wing, opposing lift, increasing the effort the bird exerts.

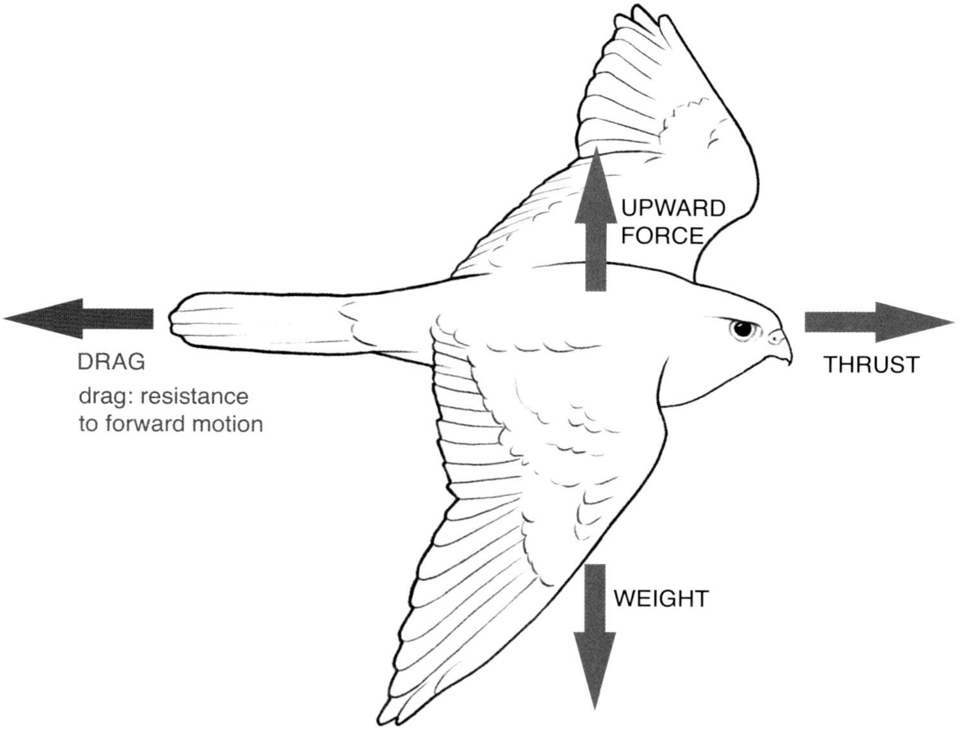

FIGURE 7.1

Forces acting on a bird in flight: drag, the resistance to forward motion. [After Alexander (2002).]

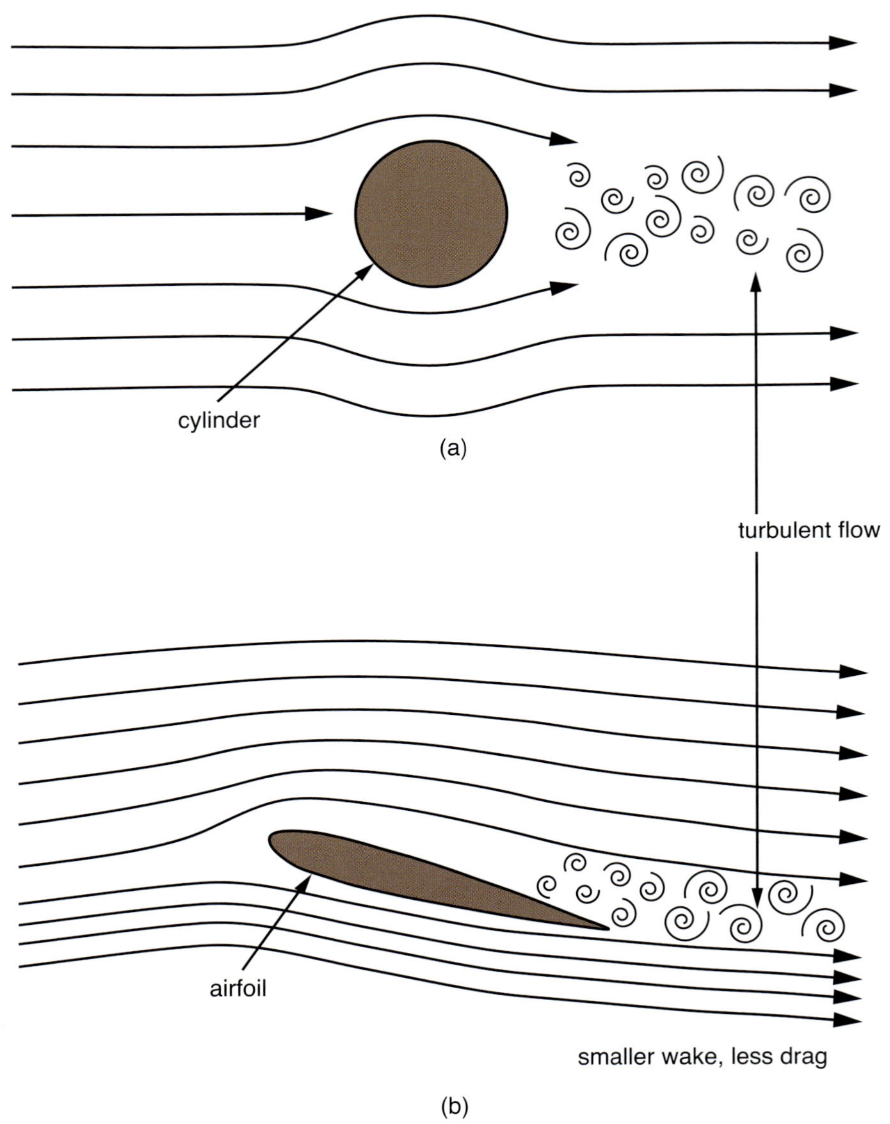

FIGURE 7.2

Pressure drag arises from the difference in pressure in front of and behind an object. [After Alexander (2002) and Shapiro (1961).]

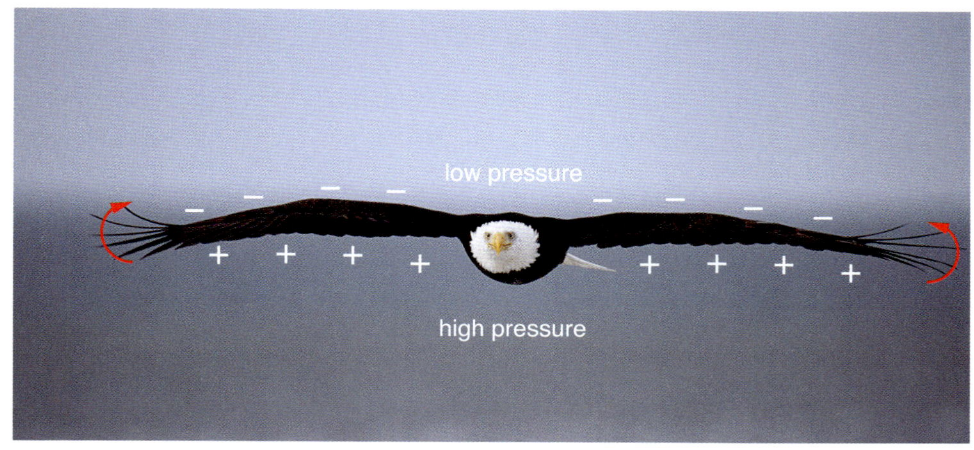

(a)

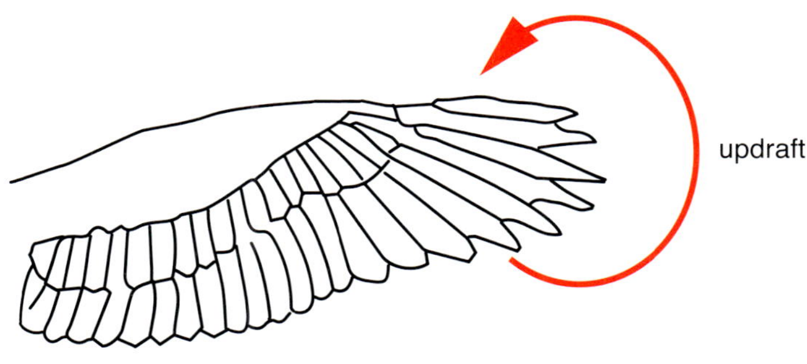

(b)

FIGURE 7.3

Wingtip vortex drag. (a) At the tip of the wing, higher pressure below the wing pushes air up and around the tip of the wing into a vortex, which then (b) pushes down on the upper side of the wing. [Photo: Alamy.]

FIGURE 7.4

Canada geese in their familiar V formation. [Alamy.]

Geese flying in their familiar V formation exploit this upward motion of air just off the wingtip (figure 7.4). Following the leader, each bird positions itself just off the wingtip of the preceding bird, getting a little additional lift from the wingtip vortex of the bird ahead of it, saving energy (figure 7.5). The birds sense when they're in the right position by feeling the additional lift (a little like a cyclist drafting behind another cyclist feeling the reduction in drag when he or she is in the right position). The lead bird expends more energy than the rest of the flock; they take turns being in the lead position.

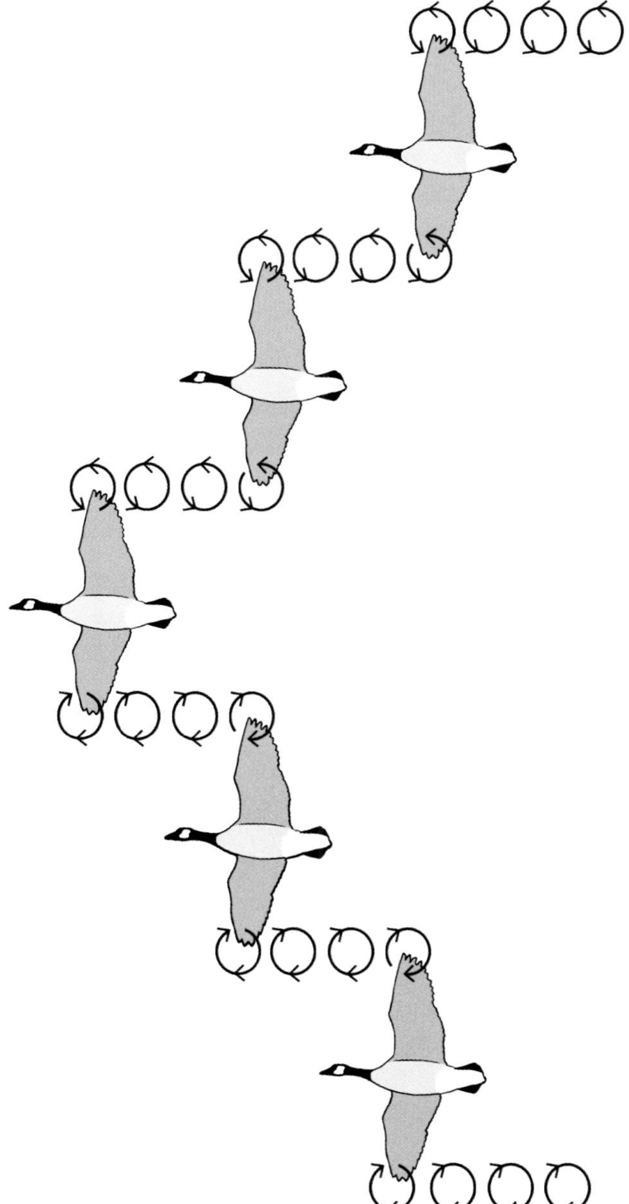

FIGURE 7.5

Canada geese in V formation. Each bird behind the leader spaces itself to get additional lift from the wingtip vortex of the bird in front of it. [After Gill (2007) and Alexander (2002).]

In one set of experiments done by Henri Weimerskirch and his colleagues, the flight of single great white pelicans and great white pelicans trained to fly in V formation were filmed with a high-speed digital camera, and the heartbeat of each bird was measured by electrodes implanted under the skin (figure 7.6). When flying alone or in the lead position of the V formation, their wings beat at about 100 times per minute, and the birds had an average heart rate of about 185 beats per minute. Birds flying in formation behind the leader spent more of their time gliding, allowing them to beat their wings less (50–60 times per minute) and lowering their heart rate to about 165 beats per minute. The birds flying in the V formation (except for the leader) were estimated to save about 10–15% of the energy expended by a bird flying alone.

In a later study, by Steven Portugal and James Usherwood at the Royal Veterinary College in London and their colleagues, a flock of northern bald ibises were fitted with GPS devices that allowed the positions of individual birds in their V formation to be tracked. The birds were also fitted with inertial devices that allowed

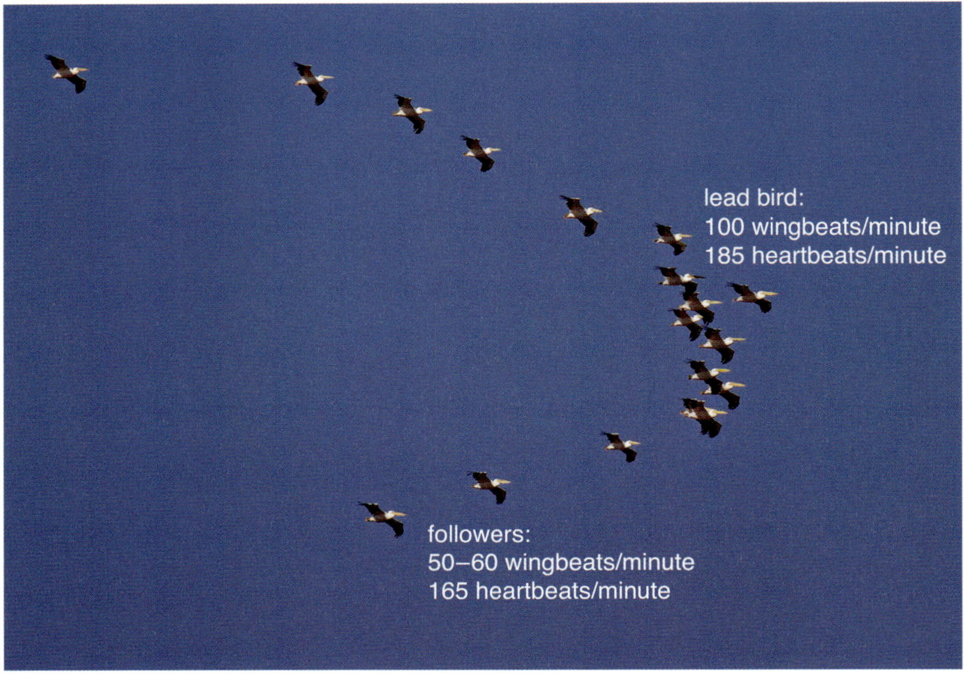

FIGURE 7.6

Great white pelicans in V formation following the leader expend about 10–15% less energy than a single bird flying alone. [Alamy.]

(a)

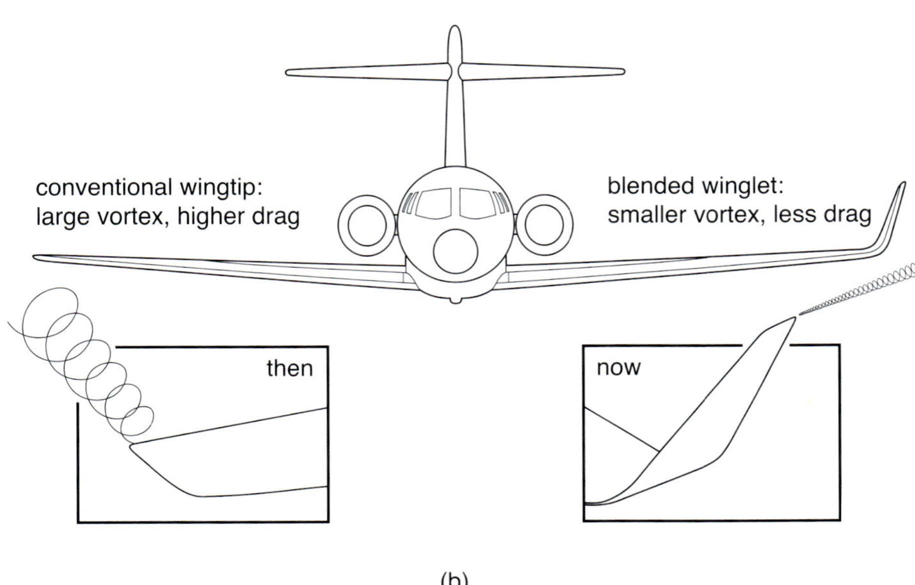

(b)

FIGURE 7.7

(a) Upturned wingtips reduce wingtip vortex drag; (b) upturned wingtips on airplanes also reduce wingtip vortex drag. [(a) Alamy; (b) after https://www.aopa.org/news-and-media/all-news/2017/may/flight-training-magazine/winglets.]

the wing flap motion of each bird to be measured. The researchers found that the wingtip position of each bird in V formation corresponded to theoretical predictions of the position that would maximize the benefit from the upward air from the vortex of the bird ahead. And not only do the birds optimize their position in the V formation, they also time the downstroke and upstroke of each flapping cycle to take advantage of the upward air from the vortex of the bird ahead.

You sometimes see birds flying with their wingtips upturned: this produces smaller wingtip vortices, reducing induced drag (figure 7.7). Modern airplanes with vertical projections at the tip of the wing (called winglets) work in the same way: the winglets reduce wingtip vortices and induced drag, reducing fuel consumption.

Wingtip vortices are what engineers call an "edge effect": they only occur at the wingtip, rather than along its entire length. For this reason, as the length of the wing increases relative to its width, the relative contribution of induced drag from wingtip vortices decreases in comparison with other sources of drag, such as pressure and skin friction drag. The larger the aspect ratio of the wing (the ratio of wingspan to mean wing width), the lower the drag and the higher the lift force relative to the drag force (figure 7.8; table 7.1). Albatrosses, among the most spectacular soaring birds, have very slender wings indeed: those of the southern Buller's albatross are 14 times as long as they are wide (figure 7.9), and those of the snowy albatross (formerly called wandering albatross) are 15 times as long as they are wide (figure 7.8b). (The snowy albatross has the largest wingspan of any living bird: up to 11.5 feet or 3.5 m.)

(a)

(b)

FIGURE 7.8

(a) Red-tailed hawk, with a wing aspect ratio of 7.1 and lift/drag ratio of 10, and (b) snowy albatross, with a wing aspect ratio of 15 and a lift/drag ratio of 19. [Both Alamy.]

FIGURE 7.9

Buller's albatross, with a wing aspect ratio of 14. [Alamy.]

Table 7.1: Comparison of wing aspect ratio for birds and planes

Flyer	$\text{Aspect Ratio} = \dfrac{\text{Wingspan}}{\text{Width}}$	$\dfrac{\text{Lift Force}}{\text{Drag Force}}$
Sparrow	5.3	4.0
Red-tailed hawk	7.1	10
Snowy albatross	15	19
Boeing 747	7	15
Standard sailplane	21	40
Competition sailplane	38	>50

From Alexander (2002).

Large, soaring birds like bald eagles sometimes spread the feathers at the tips of their wings, creating a vortex around each feather tip (figures 7.3a, 7.7a). Since each feather has a higher aspect ratio than the wing as a whole, this generates several smaller vortices at the tip of each feather instead of one larger vortex, again reducing wingtip vortex drag.

The equation for the overall drag force is similar to that for lift, with the drag coefficient, C_D, replacing the lift coefficient, C_L, and the projected area of the wing, $A_{projected}$ (the area perpendicular to the airflow), replacing the area as seen from above, A_{wing}:

$$F_{drag} = C_D \frac{\rho_{air} A_{projected} V^2}{2}$$

For bird wings as well as engineered airfoils, the drag coefficient is less than the lift coefficient; typically, it is less than half the lift coefficient. The lift force is much larger than the drag force.

SOARING AND GLIDING

Raptors such as hawks, eagles, and vultures are known for soaring and gliding. Broad-winged hawks migrate largely by soaring and gliding, from Canada and the United States to Central and South America and back again. Albatrosses soar and glide thousands of miles over ocean waves, for months on end, stopping at land only to nest. Soaring birds gain height by exploiting the interaction of air or wind currents with the land or water and then glide, sometimes for miles, almost effortlessly, simply by holding their wings out, not needing to flap. As Wilbur Wright observed, "No bird soars in a calm."

Over land, upward air currents may arise from thermals, caused by temperature differences in the land below (figure 7.10). Over hills or mountains, lower elevations are warmer than higher elevations; thermals may arise over the warmer land. The dark, brown soil of a freshly plowed field may be warmer than an adjacent patch of lush, green meadow, causing air to rise over the darker plowed field. Some thermals form in the shape of discrete rings while others form more continuous chimneys; in both cases, the upward air forms circles that soaring birds track as they gain height. The air in thermals can rise at rates of up to 13 feet per second (about 4 meters per second) and thermals can attain heights of up to about 6,600–9,800 feet (2–3 kilometers).

Soaring birds glide downward at a shallow angle, allowing them to cover large distances between thermals. By following soaring African vultures in a powered glider, Colin Pennycuick found that they traveled slightly over 46 miles (75 km) using just six thermals, each nearly a mile (1.5 km) high. On average, they flew 7.76 miles (12.5 km) between thermals.

In slope soaring, birds take advantage of the uplift from wind blowing up the slope of the mountain or ridge (figure 7.11). At Hawk Mountain, in eastern Pennsylvania, when a cold front passes through in the fall, the prevailing northwest winds are forced up Kittatinny Ridge, allowing birds to glide along the ridgetop, maintaining or even gaining altitude. Under ideal weather conditions in the fall, thousands of migrating broad-winged hawks can be seen flying along the ridge at Hawk Mountain in a single day.

What about birds that soar over the ocean, where there are no land features for the air to interact with?

Over the ocean, the wind is nearly still at the surface of the water, and its speed increases with height above the water until it reaches a roughly constant speed about 30–65 feet (10–20 m) above the water (figure 7.12). Since lift increases with the square of the air speed over the wing, a soaring seabird gets increasing lift as it rises from just above the surface of the water until it reaches wind of constant speed. Soaring seabirds like the snowy albatross gain height by flying steeply upward across the wind and then gently downward with the wind, their flight path forming asymmetric loops. A snowy albatross typically glides downward at an angle of 3°, flying forward 19 feet for every foot drop in height. (Figure 7.12 exaggerates

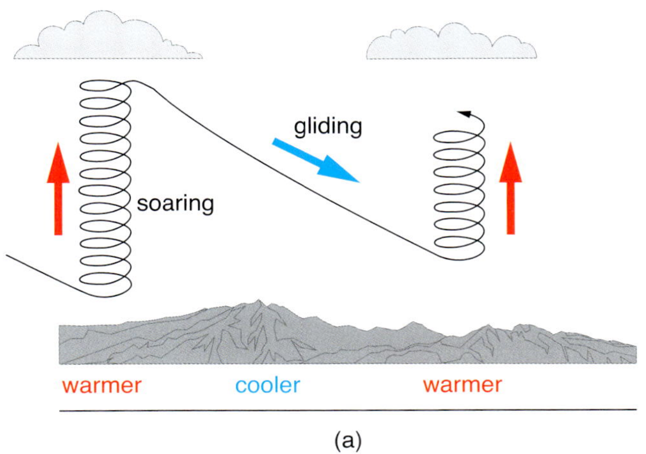

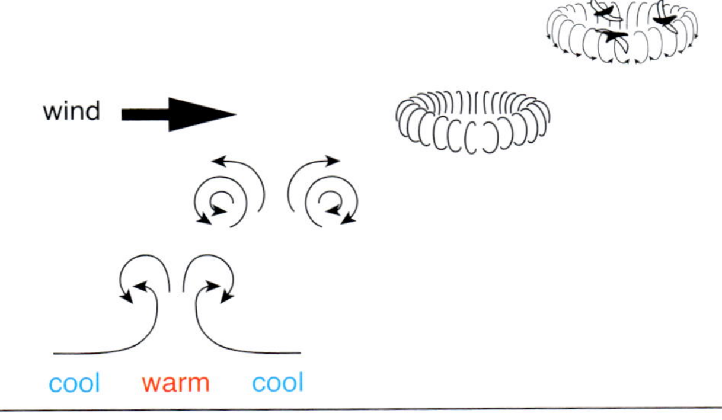

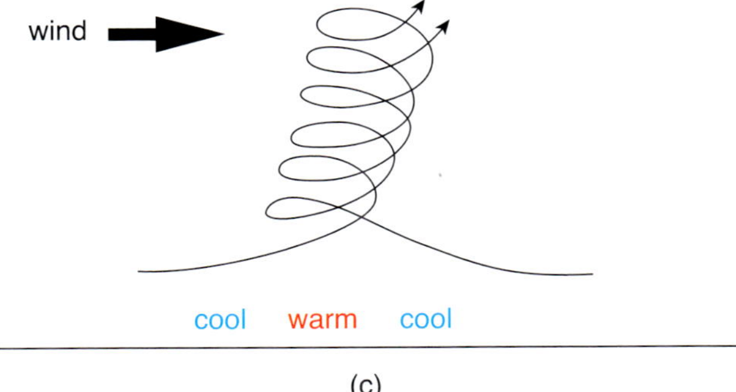

FIGURE 7.10

(a) Soaring birds rise with warm air in thermals, glide downward, and then soar up again on the next thermal. Thermals in the form of (b) discrete rings and (c) a chimney. [(a) After Gill (2007); (b, c) after Alexander (2002).]

FIGURE 7.11

In slope soaring, birds exploit winds blowing up the slopes of mountains or ridges to give them lift. [After Kerlinger (2009).]

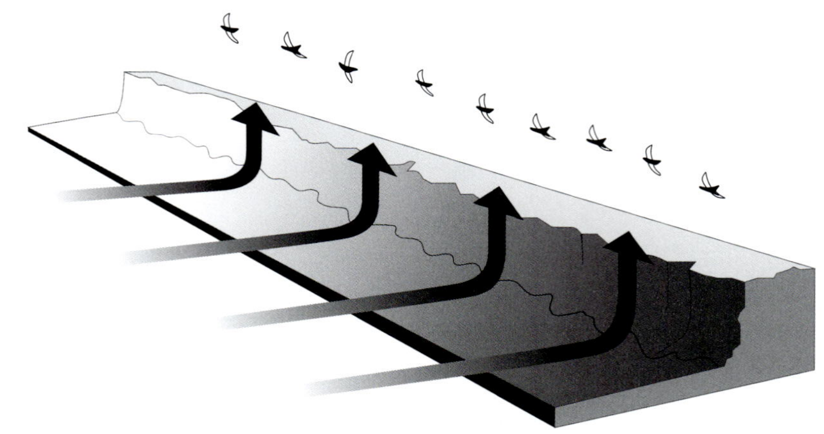

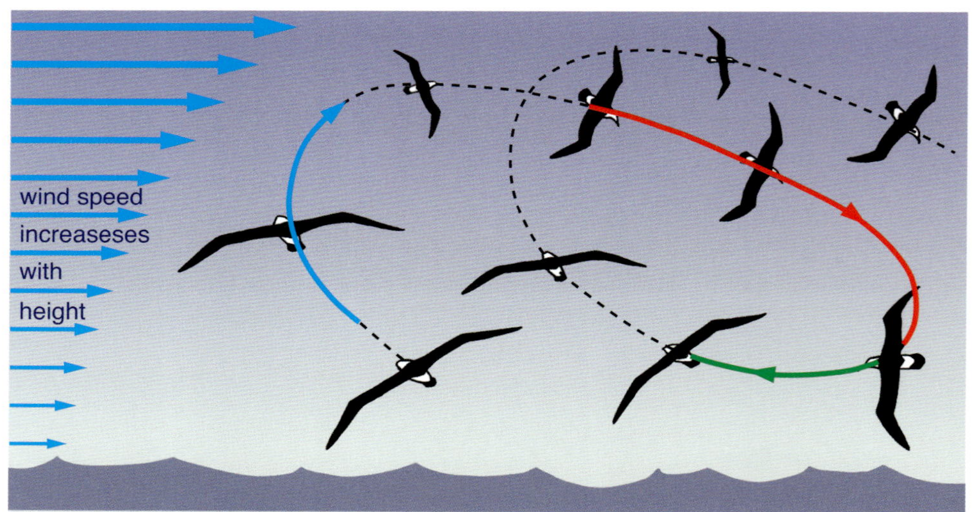

FIGURE 7.12

In dynamic soaring, a seabird gains height by flying steeply upward across the wind (blue curve) and then gently downward with the wind (red curve). As it reaches the bottom of the loop, it initially climbs using the momentum from its downward glide (green curve) before gaining further lift from the increasing speed of the wind again (blue curve). The flight path forms asymmetric loops. [After Lovette and Fitzpatrick (2016).]

the downward slope, as the figure is not wide enough to show the gentle descent realistically.) As the bird reaches the bottom of its loop and turns into the wind, it initially climbs using the momentum from its downward glide. It then climbs further by exploiting the increasing lift as the wind speed increases with height. This type of soaring is called *dynamic soaring*.

If ocean waves are sufficiently high, albatrosses (and other pelagic soarers) can exploit updrafts associated with the crest of a wave to increase their height and then glide down into a trough between waves before repeating the cycle. This is known as *wave slope soaring* (analogous to slope soaring over land). In typical conditions in the Southern Ocean, where many species of albatrosses are found, this is less important than dynamic soaring.

Although most small birds do not spend much time gliding, swallows do, especially when foraging for insects. Barn swallows, in particular, are known for foraging close to the ground, where gradients in wind speed exist, similar to those over water. Using high-speed videography along with wind measurements from an ultrasonic anemometer, Douglas Warrick and his colleagues have shown that barn swallows also exploit ground level wind speed gradients to increase their speed as they rise upward and turn over the ground.

We've already seen that African vultures can glide over 7.5 miles from the height they attain on a single thermal, and that snowy albatrosses can glide 19 feet forward for every foot drop in height, all without flapping their wings. How do they do it?

As a soaring bird glides forward, it gradually descends, moving along a glide path at a glide angle, θ (theta), to the horizontal (figure 7.13). The bird tilts its wings slightly downward, balancing the three forces we have described so far: weight (acting downward), lift (acting perpendicular to the airflow over the wing), and drag (acting parallel to the airflow over the wing). The net effect of lift, L, and drag, D, given by the resultant, R, exactly counterbalances the weight, W. The weight can be further broken down into components parallel to and perpendicular to the direction of motion (figure 7.14a). The red component parallel to the direction of motion propels the bird along its path; not only the downward motion but also the forward motion of a soaring bird is powered by gravity. The ratio of horizontal distance traveled forward to vertical distance descended is the same as the ratio of lift to drag; both ratios depend on the glide angle in the same way (figure 7.14b).

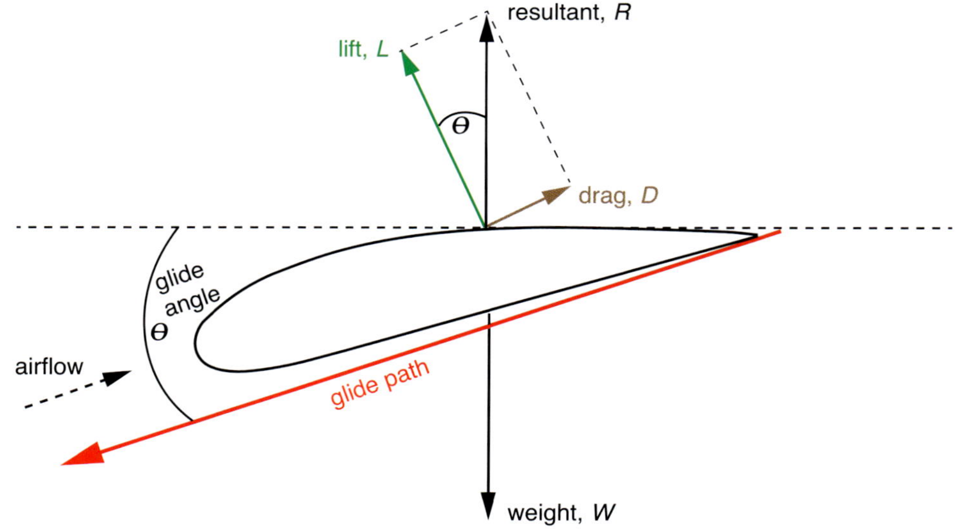

FIGURE 7.13

Forces on a wing gliding downward at a glide angle theta (θ). Lift, *L*, acts perpendicular to the airflow, and drag, *D*, acts parallel to the airflow. The combination of these two forces (the resultant, *R*) balances the weight, *W*, of the bird. [After Alexander (2002).]

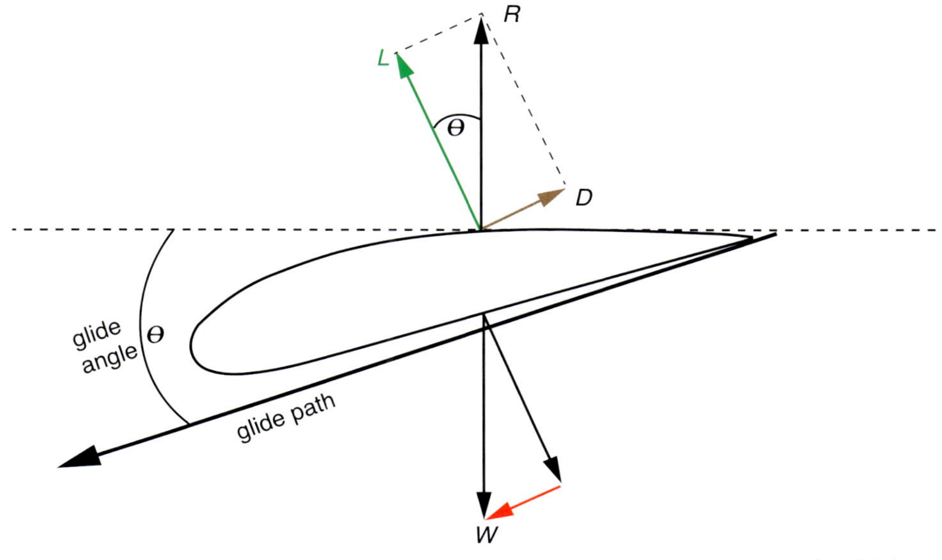

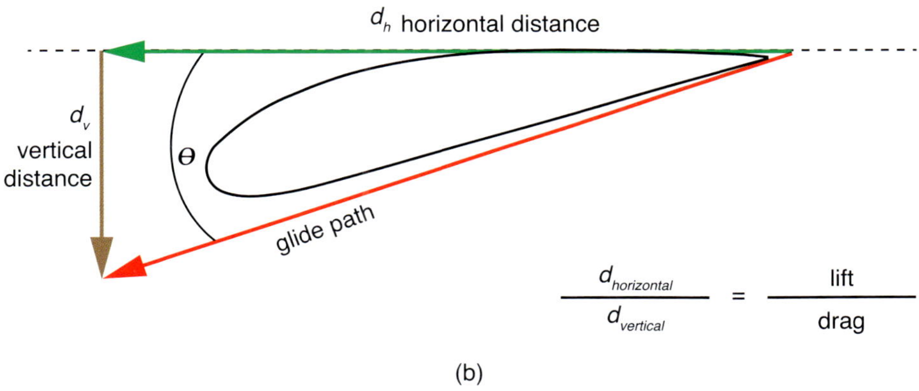

FIGURE 7.14

(a) The bird's weight can be broken down into two components; the component parallel to the glide path (red arrow) propels the bird forward. (b) The ratio of horizontal distance traveled forward to vertical distance descended is the same as the ratio of lift to drag; both ratios depend on the glide angle theta, **θ**, in the same way. [After Alexander (2002).]

Soaring birds typically have lift-to-drag ratios of about 10–20, which means that they move 10–20 feet or meters forward for every foot or meter of descent; this corresponds to glide angles of 3–6° (table 7.2). The snowy albatross has the highest ratio of forward to downward motion of all birds, 19, due to its high-aspect-ratio wings (table 7.1). For comparison, the ratio of forward to downward motion of a hang glider is typically 8, while that of a Boeing 747, if it were descending in a pure unpowered glide, is 15 (from its typical flying altitude of about 6.5 miles, a 747 can glide 98 miles). A typical sailplane has a ratio of forward to downward motion of 40: if towed to a height of one mile, it could glide 40 miles before reaching the ground.

Table 7.2: Forward/downward distance and glide angles

Flyer	Forward/downward distance (= lift/drag)	Glide angle
Red-tailed hawk	10	5.7
Turkey vulture	16	3.7
Snowy albatross	19	3.0
Hang glider	8	7.1
Boeing 747 (unpowered glide)	15	3.8
Sailplane	40	1.4

From Alexander (2002).

With this information, we can estimate how fast a soaring bird rises in a thermal. A turkey vulture, with a forward/downward distance ratio of 16, gliding at a speed of 22 mph (or 32 feet/second) horizontally over the ground, descends at a speed of 2 feet/second (32/16 = 2). The air in a thermal rises at about 13 feet/second (although this depends on the particular conditions of the thermal), so that the vulture's net upward speed in a thermal is 11 feet/second, or about a thousand feet (or 300 m) in 90 seconds. Birds soaring in thermals can gain altitude remarkably quickly.

THRUST: FLAPPING FLIGHT

How do birds generate thrust to overcome drag and produce forward motion in flapping flight (figure 7.15)?

We first need to see how the wing moves in flapping flight, and how air flows over it. High-speed video images of birds in flight reveal that on the downstroke, which produces most of the lift and forward thrust, the wing moves down and forward (figure 7.16). On the upstroke, the wingtip first moves forward and upward and then reverses and curves around to the starting position. Some birds splay their flight feathers apart on the upstroke, allowing air to pass between them, easing their return to their starting position.

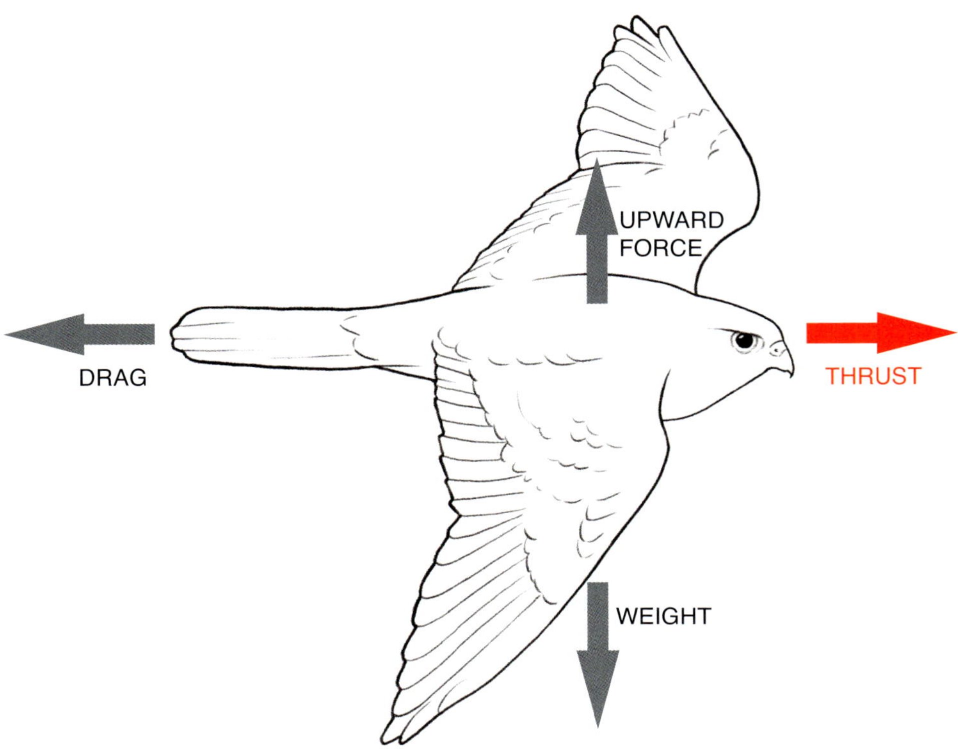

FIGURE 7.15

Forces acting on a bird in flight: thrust. [After Alexander (2002).]

FIGURE 7.16

Wing motion in flapping flight. [After Gill (2007) and Burton (1990).]

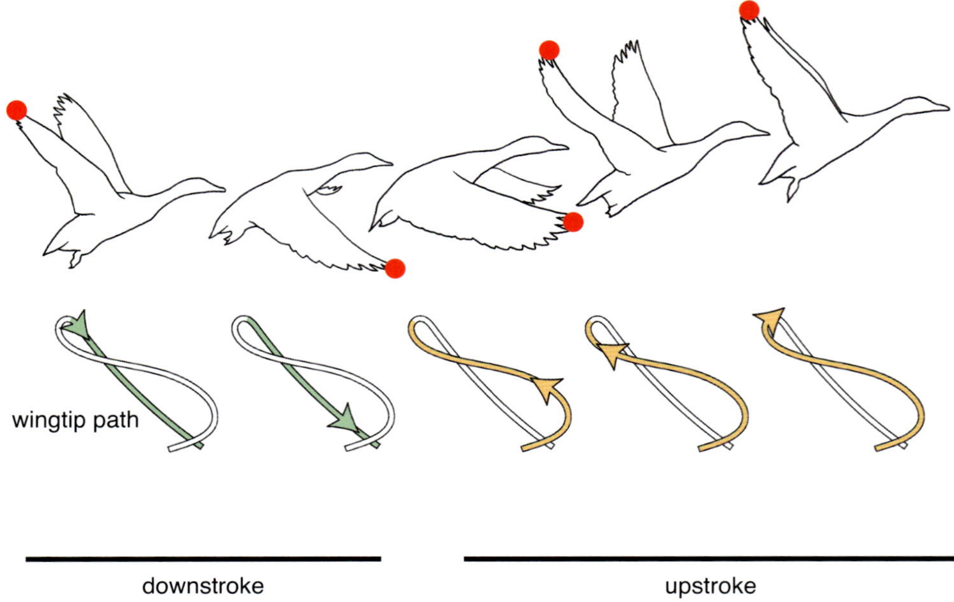

FIGURE 7.17

(a) The speed of the wing through the air is shown by the red dashed line. The airflow over the wing is in the opposite direction (black dashed line) (b) Lift, *L*, acts perpendicular to the airflow, and drag, *D*, acts parallel to the airflow; their combined effect is the resultant, *R*. (c) The horizontal component of *R* is the thrust and the vertical component of *R* is the upward force that supports the weight of the bird. [(a) After Alexander (2002); (b, c) airfoils: CC0 Olivier Cleynen, Wikipedia.]

In flapping flight, the bird has a forward speed and the wing has a downward speed; the net speed of the wing through the air is shown as a dashed red line in figure 7.17a. Relative to the bird, the speed of the air over the wing (the airflow) is in the opposite direction (figure 7.17a, dashed black line). The lift force acts perpendicular to the airflow, and the drag force acts parallel to the airflow (figure 7.17b). The combined effect of the lift and drag forces is the resultant force. The resultant force, R, has a horizontal component (figure 7.17c): this is the thrust, T, that moves the bird forward. The vertical component of R is the upward force, U, that supports the bird's weight.

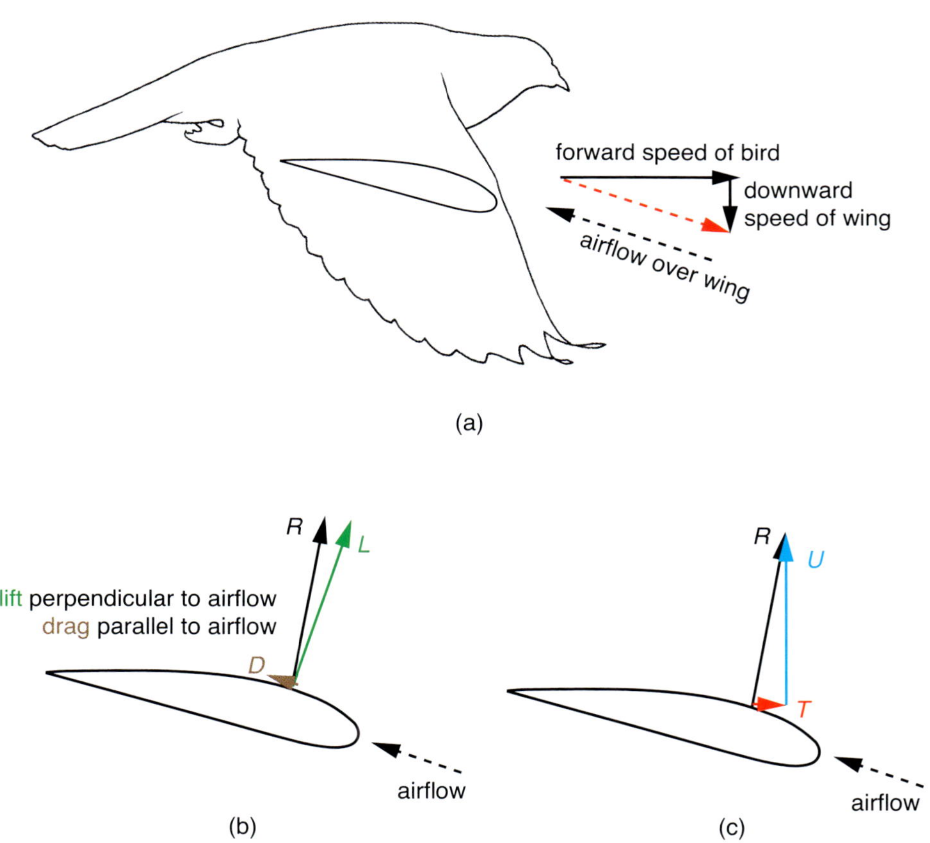

The larger the lift force relative to drag, the more the resultant *R* tilts forward and the higher the thrust (figure 7.18).

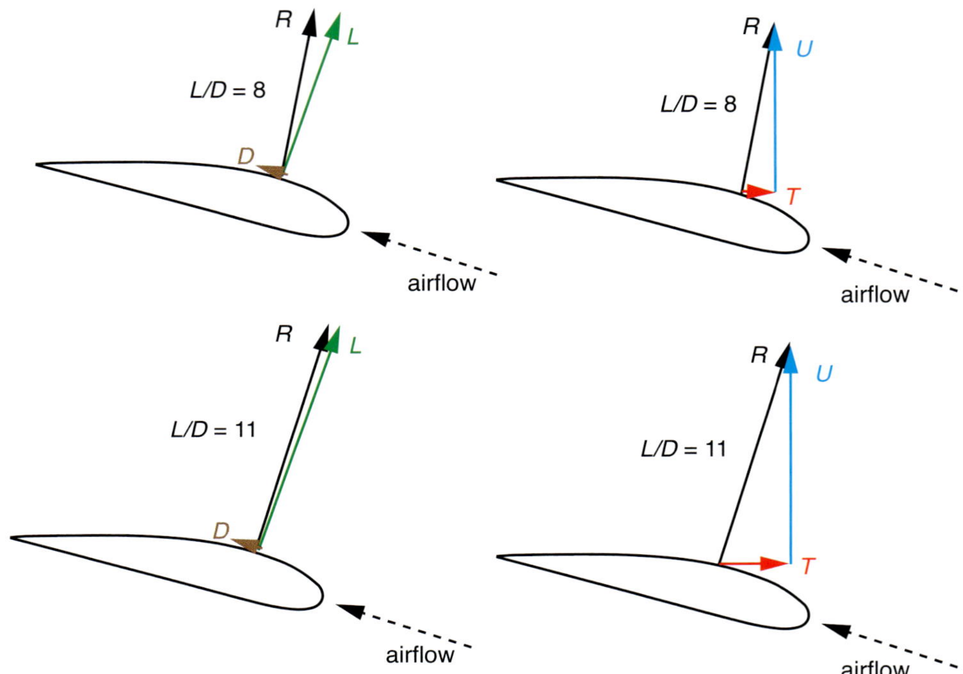

FIGURE 7.18

The larger the ratio of lift to drag, the larger the thrust. [After Alexander (2002); airfoils: CC0 Olivier Cleynen, Wikipedia.]

As a bird flaps its wings, the airflow over the wing varies along the length of the wing (figure 7.19). The forward speed is the same all along the wing (as the bird and its wings all move forward at the same speed). But the downward speed of the wing is greater at the wingtip than at the base, next to the body (since the wingtip moves through a greater distance than the base of the wing over the same time). In addition, the orientation of the cross section of the wing varies along the length of the wing: toward the wingtip, the cross section of the wing is tilted more downward than at the base.

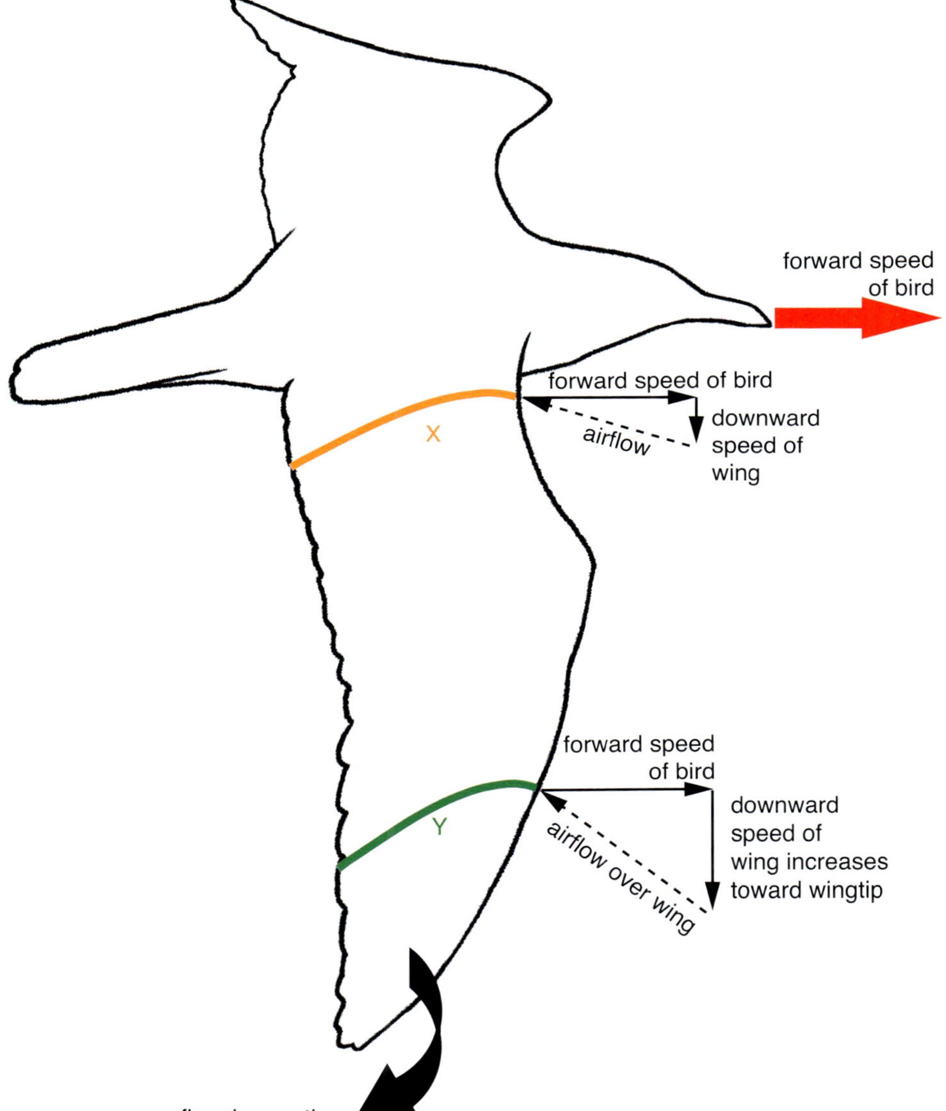

FIGURE 7.19

Airflow across the wing varies along the length of the wing. [After Alexander (2002).]

The lift, drag, and resultant forces acting at two cross sections, X, near the base of the wing, and Y, near the tip of the wing, are shown in figure 7.20. Near the base of the wing, at section X, the resultant force is close to vertical: the airflow over the base of the wing produces substantial upward force counteracting gravity, but not much thrust. Toward the wingtip, at section Y, the change in the orientation of the airflow over the wing tilts the lift force, L, as well as the resultant, R, forward: the wingtip contributes much of the forward thrust, as well as some upward force resisting gravity.

The speed of the bird increases with thrust. The faster a bird flaps its wings, the higher the downward speed of the wings, the faster the airflow over the wings, the larger the lift (and drag), the larger the resultant force, and the higher the thrust. A bird can also increase thrust and speed by increasing the angle of attack of the wing; this is limited at some point, though, by increasing drag on the wing. Data for the speed of migrating birds in level flapping flight is given in table 7.3.

Table 7.3: Speed of flight

Species	Speed (mph)	Speed (km/hr)	Speed (m/s)
Barn swallow	22	35	10
European starling	36	58	16
Arctic tern	24	39	11
Red knot	44	71	20
Peregrine falcon	26	42	12
Mallard	42	68	19
Herring gull	29	47	13
Canada goose	37	60	17
Mute swan	35	56	16

Data from Tennekes (2009).

Next, we'll take a look at a few specialized types of flight: hovering of hummingbirds; swimming of penguins and alcids; and murmurations of starlings.

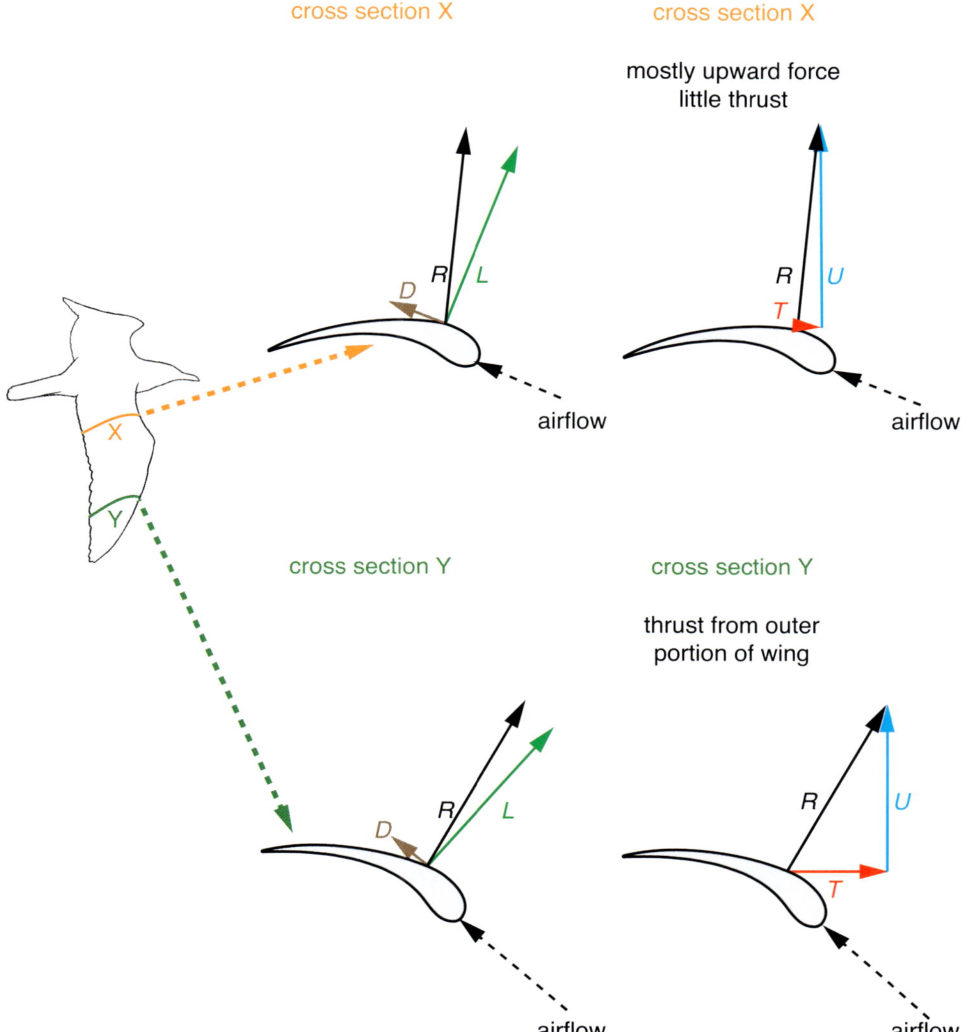

FIGURE 7.20

The lift, drag, and resultant forces acting at two cross sections, X, near the base of the wing, and Y, near the tip of the wing. Near the base of the wing, at X, the force is mostly upward, with little forward thrust. At the tip of the wing, at Y, the thrust is larger. [After Alexander (2002); airfoils: CC0 Olivier Cleynen, Wikipedia.]

SPECIALIZED TYPES OF FLIGHT

HOVERING

Hummingbirds zip from flower to flower, abruptly stopping and hovering in place to draw nectar from each one (figure 7.21). They can hover in still air and even go backward. How do they do it?

In forward flight, lift is generated by both the forward speed of the bird as well as the downward speed of the wing (figure 7.19). But when a hummingbird stops and hovers, it loses lift from forward speed and must generate an upward force to balance its weight in another way.

During hovering, a hummingbird holds its body nearly vertical, flapping its wings forward and backward, rather than the more usual up-and-down flapping of forward flight. High-speed video of hovering hummingbirds reveals that, while hovering, their wings first move forward on a roughly horizontal plane, then, at full forward extension, rotate 180°, so that the initially lower wing surface is now on top, and swing backward in nearly the same horizontal plane, so that the wing tip trajectory forms a figure-eight (figure 7.22). The wing then rotates back to the original orientation (top surface of the wing on top) for the next cycle of forward

FIGURE 7.21

An Allen's hummingbird hovering. Hummingbirds can hover even in still air. [Alamy.]

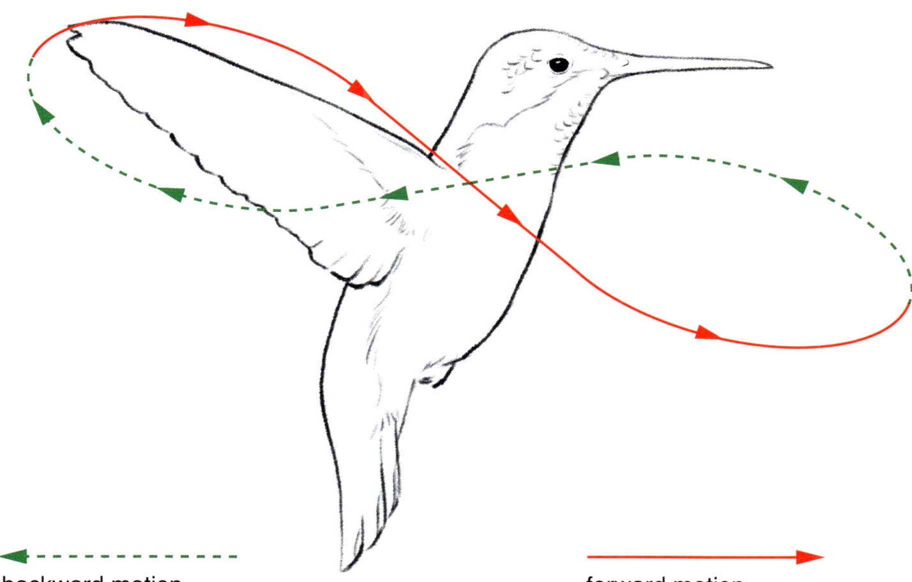

FIGURE 7.22

Hummingbird hovering. The motion of the wings is a little like that of our arms when treading water; both generate an upward force. [After Alexander (2002).]

motion. The motion of the wings is a little like the back-and-forth motion of the arms of a person treading water, with the arms rotating at the end of each stroke.

When a bird in forward flight flaps its wings up and down, it generates thrust that moves the bird forward, perpendicular to the up-and-down motion of the wings. Similarly, when a hummingbird flaps its wings back and forth, it generates a perpendicular upward force counteracting its weight (figure 7.23). Slight changes in the angle of the wings give varying amounts of upward and horizontal force, allowing hummingbirds to even fly backward (figure 7.24).

Specialized anatomical features in hummingbirds enable this unusual wing trajectory. Most notably, the articulation of the humerus at the shoulder joint allows the 180° rotation of the wing about its length, a feature not seen in other birds. In addition, using x-ray videography, Andy Biewener of Harvard University and his colleagues observed that as the wing starts to move backward (green dashed curve in figure 7.22), the inversion of the wing, with the initially lower wing surface now on top, occurs by rotation at the wrist.

Hovering hummingbirds must generate lift from the flapping of their wings, rather than from the flow of air generated by the speed of a bird flying forward, and this requires greater power. Power output of a muscle is the force generated times the distance the muscle shortens when contracted, divided by the time it takes to contract (which, for a beating wing, is proportional to the time for a single wingbeat). The force generated by the muscle is proportional to its cross-sectional area. Since the cross-sectional area of a muscle in an animal increases with the square overall length of the animal, but the weight of the animal increases with the cube of its length, as its size increases, the power output per ounce of an animal's weight decreases (for a given muscle contraction length and wingbeat frequency) (figure 7.25). For this reason, only small birds are capable of sustained hovering. Ruby-throated hummingbirds have weights in the range of 0.1–0.2 ounces (2–6 grams); for comparison, black-capped chickadees are in the range of 0.3 to 0.5 ounces (9–14 grams). The smallest hummingbird, the bee hummingbird, has a weight in the range of 0.056–0.067 ounces (1.6–1.9 grams), less than that of a US penny (2.5 grams).

What about raptors like kestrels, which seem to hover in place (figure 7.26)? True hoverers, like hummingbirds and some insects, are able to hover in still air. Kestrels and other raptors are unable to do this. They require a headwind, flying into it at exactly the same speed as the wind, so that they remain still over the ground, producing the appearance of hovering.

FIGURE 7.23

Forward flight vs. hovering. A hovering hummingbird holds its body vertical and flaps its wings roughly in a horizontal plane, generating an upward force that counteracts its weight. [Both Alamy.]

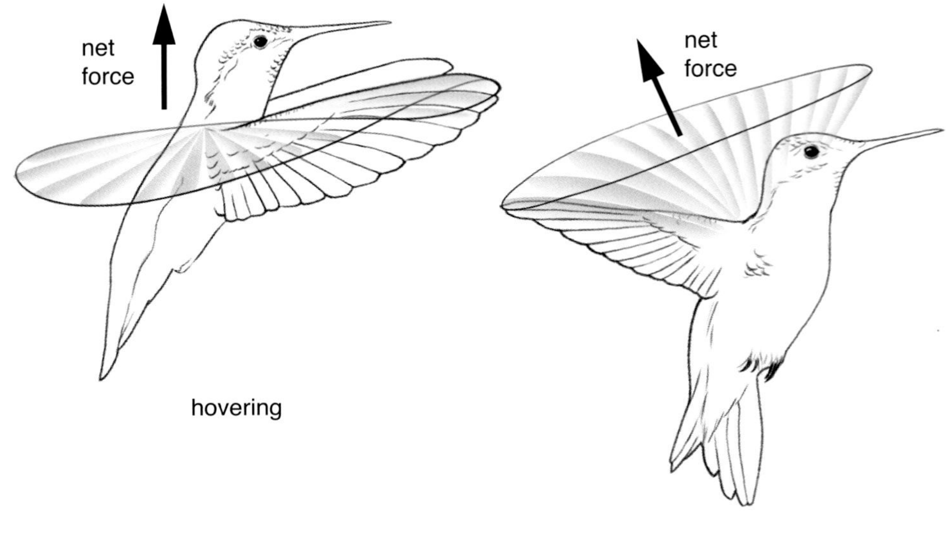

FIGURE 7.24

Slight changes in the angle of the wings give varying amounts of upward and horizontal force, allowing hummingbirds to even fly backward. [After Gill (2007) and Greenwalt (1960).]

$$\text{power} \propto \text{muscle area} \propto \text{length}^2$$

$$\text{mass} \propto \text{length}^3$$

$$\frac{\text{power}}{\text{mass}} \propto \frac{1}{\text{length}} \quad \text{decreases with increasing size}$$

FIGURE 7.25

Only small birds are capable of hovering. The power a bird can generate with its wings depends on the muscle area of the wings, which is proportional to its length squared. Its mass depends on its length cubed. With increasing size of a bird, the power it can generate per unit of its mass decreases. [Alamy.]

FIGURE 7.26

Kestrels require a headwind to "hover," flying into it at exactly the same speed as the wind, so that they remain still over the ground, producing the appearance of hovering. [Alamy.]

SWIMMING WITH WINGS

Members of the alcid family, such as puffins, murres, and guillemots, pursue prey underwater, swimming by flapping their wings (figure 7.27). They hold their inner wings close to the body and flex the wing at the "wrist," extending the "hand" and its primary feathers. As we saw in the section on thrust, the inner part of the wing produces mostly upward force and little thrust. When swimming, the alcids don't need upward force due to their buoyancy in water, so the inner part of the wing can be held close to the body to reduce drag, as can be seen in figure 7.27. To swim forward, they need thrust, which is produced mostly by the outer part of the wing.

FIGURE 7.27

A common murre (also called a common guillemot in Eurasia) swimming with its wings. All alcids, such as murres, puffins, razorbills, and guillemots, swim with their wings. Penguins, too, swim with their wings. [Alamy.]

(a)

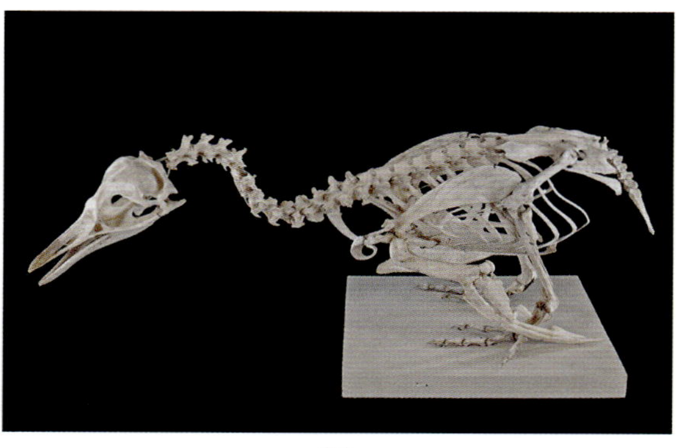

(b)

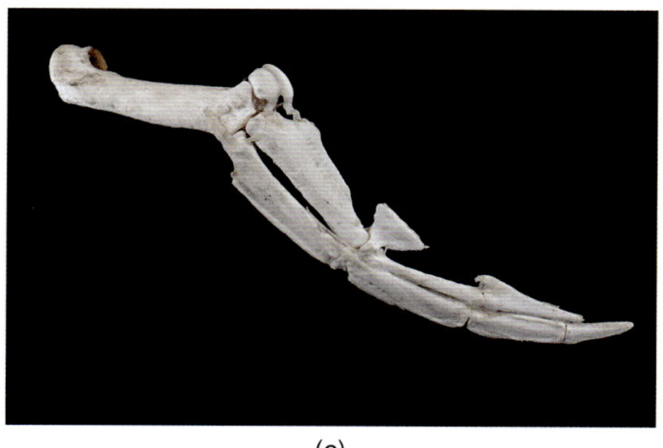

(c)

Penguins, too, swim underwater, using their short, stubby wings in a manner that corresponds somewhat to how a seal uses its flippers. The cross sections of their wing bones are flattened and elliptical, rather than circular, providing more support over the width of the wing (figure 7.28). Their fused hand bones form an arc with the radius and ulna, giving the wing a smooth shape, reducing drag through the water. While most penguins swim at 4–7 mph (6–11 km/hr), some can swim surprisingly fast: gentoo and Galapagos penguins can reach speeds of up to 22 mph (35 km/hr).[1] While snorkeling on a trip to the Galapagos Islands, I was completely startled to see a Galapagos penguin zoom past me.

FIGURE 7.28

(a) A pair of African penguins, (b) the skeleton of an African penguin, and (c) the wing bones of an African penguin. The fused hand bones form an arc with the radius and ulna, giving the wing a smooth shape, reducing drag through the water. [(a) Alamy; (b, c) skeleton: Museum of Comparative Zoology, Harvard University, Ornithology Department, specimen 341353, © President and Fellows of Harvard College.]

MURMURATIONS OF STARLINGS

The swirling, undulating flight of huge flocks of European starlings in the early evening, prior to roosting, is mesmerizing (figure 7.29). Murmuration, the onomatopoeic name for this, comes from the sound of the flock's wingbeats. Murmurating flocks often involve thousands (up to hundreds of thousands) of starlings, all moving in a collective swarm. While there isn't a complete explanation of how they do it, there have been detailed studies that give some insight into the collective motion of the flock.

One study, by Anne Goodenough, of the University of Gloucestershire, and her collaborators, gathered data on the date, location, size, and duration of over 3,000 murmurations, based on surveys from citizen scientists, mostly from the United Kingdom. The citizen scientists also reported whether or not birds of prey were present near the flock. (I love the paper's title: "Birds of a feather flock together.") The average size of the flock was over 30,000 birds; the largest flock observed was estimated to have a stunning 750,000 starlings. The average size of the flocks increased from October to early February, then decreased until March, when the season for murmurations ends. The average duration of the murmuration was just under half an hour, with little variation. At nearly one-third of the murmurations, birds of prey were present near the flock, suggesting that murmurations are a response to the presence of predators.

Another study by a group of scientists led by Andrea Cavagna, at the Institute for Complex Systems of the National Research Council in Rome, and Giorgio Parisi, at the Sapienza University of Rome, took advantage of a well-known starling roost at the Rome Termini railway station. Three video cameras, known distances apart, were set up 100 feet (30 meters) above ground level on the terrace of the Museo Nazionale Romano's Palazzo Massimo, a little less than 300 yards (250 meters) from the train station. The cameras took synchronized video images, at 10 frames per second, of the murmurating starlings over a period of 8 seconds. Using stereophotography of the three different views, along with computer analysis of the images, the researchers were able to reconstruct, in three dimensions, the position of each bird in the flock as they flew (figure 7.30). They limited their analysis to video sequences in which the birds flew in view of all the cameras and the flock was within 250 meters of the cameras; their data included flocks of up to 2,600 birds. While the researchers did see larger flocks, comprised of tens of thousands of birds, in those cases the birds were too far away to effectively use the stereophotography technique. From their analysis, they reported the shape, size, and movement of the flock as well

FIGURE 7.29

Murmurations of starlings. [Both Alamy.]

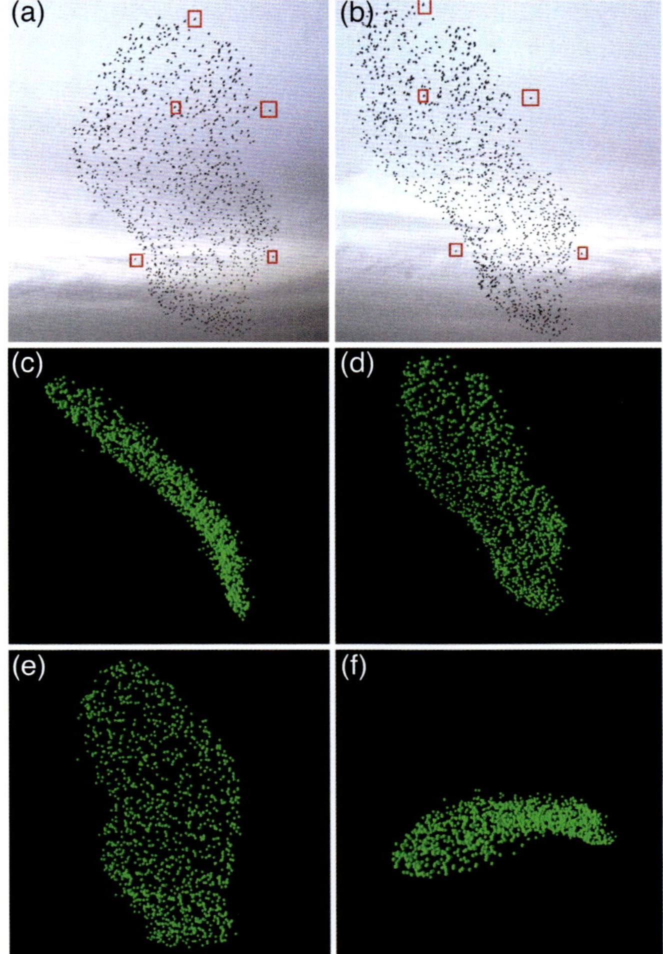

FIGURE 7.30

Tracking a murmuration of starlings. (a, b) Stereo pair images taken at the same time, 82 feet (25 m) apart; five birds are highlighted by red boxes. (b) and (d) are the same perspective; (c), (e) and (f) are from different perspectives. The imaging allowed them to track individual birds in the flock. [From Ballerini et al. (2008a), © Copyright (2008) National Academy of Sciences, U.S.A.]

as the number of birds per unit volume of space and the distribution of the birds. Their results are a remarkable feat of image processing.

The longest dimension of the flock was typically parallel to the ground, and most of the time the flock flew horizontally, parallel to the ground. The flocks were relatively thin: the vertical thicknesses were between 16 and 65 feet (5–20 meters), while the width and length were, on average, 2.8 and 5.6 times the thickness, respectively. The reason for the thin flock, generally flying roughly parallel to the ground, may be gravity: it requires more energy for a bird to increase or decrease its altitude, compared with moving sideways at one altitude. And modeling has shown that there is less drag in flocks that extend laterally rather than vertically.

The average distance between nearest neighbors was between 1.7 and 3.8 times the average wingspan of a starling and did not depend on the size of the flock. The starlings maintained a minimum distance between each bird of just under one wingspan, to avoid collisions; this distance did not depend on the size or density of the flock. The flocks were found to be denser, with birds closer together, at the edges of the flock than in the interior, possibly to deter predation; the more isolated a starling is, the more susceptible it is to attack by a bird of prey.

Further modeling and analysis have shown that each starling in the flock interacts with the six or seven birds that are its nearest neighbors and that this helps maintain the cohesion of the flock, reducing the likelihood of smaller groups of birds breaking off from the main flock in the presence of a bird of prey.

When I'm out early walking my old labradoodle, Maddie, at Jamaica Pond, enjoying the golden glow in the treetops across the pond as the sun rises, I often find myself taking in the wonders of flight.

On one outing, the morning calm is disturbed by the sudden honking of a group of 20 or so Canada geese at the edge of the pond. With some commotion and splashing, they take off over the water, rise up, and turn to head over the trees. Within a minute or so, they assemble themselves into the classic, evenly spaced out V formation. As I watch them fly off, I think of how each trailing bird catches a bit of an updraft from the vortex formed at the wingtip of the one ahead of it, making its flight a bit easier.

As Maddie and I continue our walk, a few hooded mergansers take off, flying remarkably fast, flapping their wings furiously, keeping low over the water. I think of the way the air curving over their wings produces lift and how the motion of each wingbeat thrusts them forward.

Later on in the afternoon, at the Arnold Arboretum I see a red-tailed hawk circling, gaining height on an updraft, then gliding effortlessly in its search for an unsuspecting squirrel or snake. I think of how, as the hawk glides, gravity draws it downward but also propels it forward, so that all it needs to do to stay aloft is to keep its wings outstretched.

Learning the secret of flight from a bird was a good deal like learning the secret of magic from a magician. After you once know the trick and know what to look for, you see things that you did not notice when you did not know exactly what to look for.

Orville Wright

REFERENCES

Alexander DE (2002). *Nature's Flyers: Birds, Insects and the Biomechanics of Flight*. Johns Hopkins University Press.

Babinsky H (2003). How do wings work? *Physics Education* **38**, 497–503.

Ballerini M, Cabibbo N, Candelier R, Cavagna A, Cisbani E, Giardina I, Lecomte V, Orlandi A, Parisi G, Procaccini A, Viale M, and Zdravkovic V (2008a). Interaction ruling animal collective behavior depends on topological rather than metric distance: evidence from a field study. *PNAS* **105**, 1232–1237.

Ballerini M, Cabibbo N, Candelier R, Cavagna A, Cisbani E, Giardina I, Orlandi A, Parisi G, Procaccini A, Viale M, and Zdravkovic V (2008b). Empirical investigation of starling flocks: a benchmark study in collective animal behaviour. *Animal Behaviour* **76**, 201–215.

Burton R (1990). *Bird Flight: An Illustrated Study of Birds' Aerial Mastery*. Facts on File.

Catry P, Phillips RA, and Croxall JP (2004). Sustained fast travel by a gray-headed albatross (*Thalassarche chrysostoma*) riding an Antarctic storm. *Auk* **121**, 1208–1213.

Cavanagh P (2024). *How Birds Fly: The Science and Art of Avian Flight*. Firefly Books.

Flindt R (2006). *Amazing Numbers in Biology*. Translated by Neil Solomon. Springer.

Gill FB (2007). *Ornithology*. Third edition. WH Freeman.

Goodenough AE, Little N, Carpenter WS, and Hart AG (2017). Birds of a feather flock together: insights into starling murmuration behavior revealed using citizen science. *PLOS One* **12**, e0179277.

Greenwalt CH (1960). *Hummingbirds*. Doubleday.

Hedrick TL, Tobalske BW, Ros IG, Warrick DR, and Biewener AA (2012). Morphological and kinematic basis of the hummingbird flight stroke: scaling of flight muscle transmission ratio. *Proc. Roy. Soc.* **B279**, 1986–1992.

Kerlinger P (2009). *How Birds Migrate*. Second edition. Stackpole Books.

Lovette IJ and Fitzpatrick JW, eds. (2016). *The Cornell Lab of Ornithology Handbook of Bird Biology*. Third edition. Wiley.

McCullough D (2015). *The Wright Brothers*. Simon and Schuster.

Portugal SJ, Hubel TY, Fritz J, Heese S, Trobe D, Voelkl B, Hailes S, Wilson AM, and Usherwood JR (2014). Upwash exploitation and downwash avoidance by flap phasing in ibis formation flight. *Nature* **505**, 399–404.

Richardson PL (2010). How do albatrosses fly around the world without flapping their wings? *Progress in Oceanography* **88**, 46–58.

Roberson JA and Crowe CT (1975). *Engineering Fluid Mechanics*. Houghton Mifflin.

Sachs G, Traugott J, Nesterova AP, Dell'Omo G, Kummeth F, Heidrich W, Vyssotski AL, and Bonadonna F (2012). Flying at no mechanical energy cost: disclosing the secret of Wandering Albatrosses. *PLOS One* 7, e41449.

Shapiro AH (1961). *Shape and Flow: The Fluid Dynamics of Drag*. Heinemann.

Tennekes H (2009). *The Simple Science of Flight: From Insects to Jumbo Jets*. Second edition. MIT Press.

Warrick DR, Hedrick TL, Biewener AA, Crandell KE, and Tobalske BW (2016). Foraging at the edge of the world: low-altitude, high-speed manoeuvering in barn swallows. *Phil. Trans. Roy. Soc.* **B371**, 2015.0391.

Watanuki Y, Wanless S, Harris M, Lovvorn JR, Miyazaki M, Tanaka H, and Sato K (2006). Swim speed and stroke patterns in wing-propelled divers: a comparison among alcids and a penguin. *J. Exp. Biol.* **209**, 1217–1230.

Weimerskirch H, Martin J, Clerquin Y, Alexandre P, and Jiraskova S (2001). Energy saving in flight formation—Pelicans flying in a "V" can glide for extended periods using other birds' airstreams. *Nature* **413**, 697–698.

Wright, O. Quoted in Malsbury E (2020). How we lifted flight from bird evolution. *Smithsonian Magazine*, December 17.

Wright, W. Quoted in McCullough (2015), p36, 52. [Wilbur Wright's letter to the Smithsonian is available at the Smithsonian archives, https://siarchives.si.edu/collections/siris_arc_393141.]

EPILOGUE: SANCTUARY AND SOLACE

Where are all the books for a general audience written by engineers? Looking around the science/nature sections of bookstores over many years, I saw plenty of books written by biologists, mathematicians, and physicists. But hardly any by engineers. I had a pretty good idea of what's involved in writing a book, having coauthored several academic books. And I'm a birder: I thought that perhaps birds would be a good vehicle for such a book.

I had already read a number of wonderful popular books about birds—Thor Hanson's *Feathers: The Evolution of a Natural Miracle*, Tim Birkhead's *The Most Perfect Thing: Inside (and Outside) a Bird's Egg*, Scott Weidensaul's *Living on the Wind: Across the Hemisphere with Migratory Birds*, and more. I thought I could add something to the available books by explaining how birds work using ideas I had learned as an engineer.

Reading ornithology textbooks and research articles in scientific journals, I found myself amazed, over and over again, at how birds work. The tiny, microscopic features in feathers that make them iridescent. The internal structures of bones that make them lightweight. The remarkable fringes in hummingbirds' tongues that unfurl and furl to collect nectar. The pressure-sensing cells in sanderlings' bills that allow them to detect prey beneath the sand without touching it. The shape of eggs being related to a species' flight ability and aerodynamic body shape. And how birds fly. All of it fascinating. It was wonderful, and I loved it. I began writing in earnest in 2016 and by the spring of 2019 had a draft of this entire book.

I started giving talks based on the book, some to birders and others to academics, sometimes very similar talks with different titles (for birders, "Fantastic Feathers: Form and Function"; for academics, "Feathers: Hierarchical Structure and Multifunctionality"). Both audiences loved the talks. At National Audubon's bird camp at Hog Island, Maine, one audience member commented that my talk was the highlight of the week—and they had seen puffins! Another summer, a teenage birder came up to me after the talk and said, "I take science in high school, but I didn't know you could do *this* with it!" I loved doing the talks, too.

And then everything changed.

My wife, Jeannie, had for a while been noticing a tremor in her hands, but this didn't seem like a big deal to me. She also worried about forgetting words (what our friend Lorie Howley calls "noun disease"), but she was nearly 70, and this didn't seem so different from any of our friends our age. Wanting to get it checked out, she went to see her physician's assistant, who sent her for a brain MRI. We got the phone call the next evening: she had a brain tumor. A biopsy the following week, in early April, revealed glioblastoma, the most aggressive kind of brain tumor.

It just didn't seem real. Jeannie was so healthy. In her forties she had biked across the country, not once but twice, first from Anacortes, Washington to Bar Harbor, Maine, and a few years later from San Diego, California to St. Augustine, Florida. She had done both trips fully "loaded"—carrying all her camping and other gear. And she continued to bike as she got older, commuting to Harvard Medical School, where she worked connecting students with community-service projects; riding around the Boston area; and touring with me—most recently from Pittsburgh to Washington, DC. She loved hiking, especially trips to the national parks in Utah, with their red, red rocks. She was especially proud of hiking Angel's Landing, in Zion National Park, a narrow trail bordered by sheer cliffs on either side. She did yoga and tai chi. She was careful about what she ate.

We met at a neighborhood holiday party at the end of 1997, introduced by a mutual friend. Our first date we called the glass date. She wanted to show me the Mapparium at the Mary Baker Eddy Library, which I didn't know about: a gigantic, three-story glass globe, with the countries depicted in stained glass, in brilliant reds, yellows, oranges, greens, and purples. Even better, the Mapparium has an internal walkway across its diameter, so that you can look from the inside out at the world (at least as it was when the Mapparium was completed in 1935). On top of that, the Mapparium is a remarkable whispering gallery: whispers can be heard across it, with the sound reflecting off the globe's glass surface. I wanted to show her the glass flowers at the Harvard Museum of Natural History, which she had never heard of. The glass flowers are remarkably accurate models of flowers and their interior structures, crafted by Leopold Blaschka and his son Rudolf in the late 1800s and early 1900s, commissioned by Harvard Professor George Lincoln Goodale for use in his botany lectures. The first time I saw them, I thought that they were real and kept looking around for the glass flowers. She loved the glass flowers, too. We laughed at the quirkiness of our glass visits and made plans to see each other again.

We bonded over shared interests. Luckily, I had also gone on bike trips on my own, although shorter than Jeannie's. My version of bike touring involved choosing some scenic region of France, renting a bike, and pedaling my way from one town to the next, phoning ahead to the next town each afternoon to find a two-star hotel to stay in. My first bike trip was along the Loire Valley, with its chateaux and hillside vineyards. Another year I biked around the north of France, visiting Gothic cathedrals, in preparation for a first-year seminar I was giving in the fall on how the forms of structures developed over the centuries, depending on the materials used to build them. Most recently, I had biked along the coast of Normandy and Brittany. I remember telling Jeannie that I would bike with her across a small, flat European country like the Netherlands; we eventually did do a bike tour there. We both loved walking, too, exploring trails around Boston and beyond.

That first summer, we traveled to Europe. I was shocked that Jeannie didn't have a passport and had never been out of the United States—I had been traveling to England since I was four. Passport obtained, we went to art galleries in London, visited my old friends in Cambridge, where I had been a graduate student, and had a few wonderful days in Paris, seeing the sights and developing a taste for crème brûlée. We then took a train to Orléans to start biking west along the Loire Valley. It was perfect—beautiful scenery, historic chateaux, the occasional wine tasting, biking, eating the delicious French food. The trip sealed the deal: by the time we got back to Boston, we were talking about moving in together.

We settled into sharing our lives together, going for weekend bike rides, having friends over for dinner, gardening, traveling, catching up at the end of each day. Jeannie worked with students at the Harvard Medical School on community-service projects—helping students start an HIV testing and counseling service, connecting them with local nonprofits, sometimes just listening to their stories and problems.

We married in May of 2004, as soon as gay marriage became legal in Massachusetts. She retired in 2013 and had several great years before her diagnosis, going on fall hiking trips to Utah, and lots of trips with me—summer hiking in the Tetons, winter trips to Sedona, two-week stays at a cabin in the Adirondacks that we both loved, biking, hiking, birding, just relaxing by the pond there. She trained our labradoodle, Maddie, to be a therapy dog, and took her to assisted-living residences in our neighborhood. She volunteered at LivableStreets Alliance, a local nonprofit that promotes walking, biking, and public transit in the Boston area. She loved seeing old friends and making new ones. One of her oldest friends said she had a genius for friendship. And throughout, love always.

After her diagnosis, Jeannie underwent the standard treatment for glioblastoma—six weeks of radiation, five days a week, combined with an oral chemotherapy drug that she took at home. We hoped this would buy us more time. The first few weeks, we often biked or walked the two miles between our home in Jamaica Plain and the Dana Farber Cancer Institute, going along the string of parks and ponds of Boston's Emerald Necklace.

We had two mottos: "Enjoy every day" and "Now is not the time to hold back." We were able to spend two weeks that summer in the Adirondacks, near Saranac Lake, staying at the cabin we loved. It's an idyllic location: on a pond, in the woods, with only a few other cabins nearby. During our stays, we often saw bald eagles perched in a pine tree across the pond, or circling overhead, loons diving and feeding their young, a female common merganser swimming along, trailed by over a dozen juveniles, and cedar waxwings darting out from a tall pine, over the water, catching insects and returning to perch on a branch. One early morning I watched in delight as an otter dove and surfaced over and over. The area is perfect for biking and hiking—rolling hills, some smallish mountains, beautiful vistas of the High Peaks to the south. We loved it all. That summer, we managed short bike rides and hikes in the mornings, when she had more energy, and spent afternoons sitting on the dock just enjoying the view, sometimes watching a couple at another dock, further along the shore, throw stick after stick into the pond for their three golden retrievers (who, they said, were on their "dog holiday").

After we returned to Boston, we went for shorter walks and then in the fall, as Jeannie got weaker, car rides through the Arboretum and up the hill at Larz Anderson Park with its great views of the city. Still enjoying being outside in nature. By mid-November, she was in home hospice, no longer leaving the house. On my walks, I would take photos of interesting things I spotted to show Jeannie: the pattern of fallen red and gold leaves on the sidewalk, a red-tailed hawk on a lamp post, snow on a holly bush between its waxy green leaves and red berries, snow draped over the branches of the conifers at the Arboretum. I told her we won the love lottery.

That fall and winter, I saw a social worker, Bobbi Allison, at the Dana Farber regularly. Most times I would walk along the Emerald Necklace to get there, bringing my binoculars to do a little birding on the way. There was always something to see: hooded mergansers at Jamaica Pond, common mergansers at Leverett Pond, a kingfisher cackling from a branch overhanging Leverett Pond, wood ducks in the little stream along Brookline Avenue between Leverett Pond and the Muddy River. I often started our sessions with my bird report, showing Bobbi photos from the Cornell Lab of Ornithology of the birds I'd seen on my walk over. It always amused me that I was the only one walking around the Dana Farber with binoculars.

We were fortunate in many ways—Jeannie remained good-natured and wasn't anxious at all, perhaps due to the way the tumor affected her brain. She just got progressively weaker. She wasn't in pain until near the end, when we were able to control it with morphine. Many friends and family visited and stayed in touch. Our wonderful friend Dale, a home health aide, came to work for us four days a week, taking care of both Jeannie and me. We were financially secure enough to be able to afford help. And MIT was generous in giving me time off to be with Jeannie. We were grateful to everyone who helped us throughout.

On March 11, 2020, Jeannie passed away. For several weeks before, knowing that she wasn't going to last a lot longer, it felt like there was a great weight hanging over me that was about to drop. But there was another weight about to drop, one that I hadn't bargained for: the same day that Jeannie died, the World Health Organization declared the coronavirus outbreak a global pandemic.

To say that the next few months were difficult would be an understatement. There was so much uncertainty and anxiety about the virus. It was difficult to see people. It wasn't possible for my family in Canada to visit me in Boston or for me to travel there. The only way I could gather family and friends for a memorial service was on Zoom. But I told myself to just keep going, just keep going.

And what kept me going were two things: staying in touch with friends and family, even if just by phone and Zoom, and going out for walks. A *lot* of walks. On the lookout for birds, or anything interesting happening in the natural world. Along the Emerald Necklace, around Jamaica Pond, through the Arnold Arboretum, to the Massachusetts Audubon Society's sanctuaries, to the Blue Hills, along the beach at Wellfleet on Cape Cod. And even on the hardest days, when missing her felt like a physical pain, I would be out on a walk, spot something—the kingfisher cackling from its perch on a branch overhanging Leverett Pond, or a wood duck paddling on Jamaica Pond or a hawk circling overhead—and still stop in awe and think: *Oh, wow, I love seeing that*. And for that moment, the grief would disappear.

A few months after she died, I started writing a *Nature Notes* blog, just for friends and family, of what I saw on my walks, taking photos with my cell phone. I saw some amazing things right near my house. Two northern flickers, perched on a high tree branch just out my back door, facing each other, swerving and jabbing their bills into the air toward each other, in what the Cornell Lab of Ornithology *All About Birds* website calls a "fencing duel." An albino squirrel on a grassy area by Jamaica Pond. A turtle laying spherical, white, ping-pong-ball-like eggs, one after another, in a hole she had dug, right by the Longwood trolley station, the stop for the big Harvard teaching hospitals, seemingly unconcerned by the commuters walking by, stopping to watch. A female wood duck in the Muddy River with nine ducklings. Throughout all of this, nature was a sanctuary and solace that helped me get through my grief.

In the fall I went to the Adirondacks, to the cabin we used to rent, to scatter Jeannie's ashes into the pond. Early the next morning, just after sunrise, when I opened the curtains of the cabin window, with the light still at that stage of diffuse pinkish glow, a bald eagle flew past right at eye height, above the steep hillside down to the water, less than 50 feet away, flashing by in seconds. Made me catch my breath in wonder. Even though I'm not a religious person, it felt almost like a sign from Jeannie.

I went back to teaching my MIT courses in the fall of 2020 and spring of 2021, all on Zoom. I had already put my classes online, so I was fortunate: I already had professionally produced lecture videos, edited into 5- or 10-minute segments, and problem sets online. As I thought about how to teach that year, I realized that what the students really wanted was contact with me, so I arranged a regular lecture schedule, and we watched the lecture videos together. The breaks between segments gave the students time for discussion and questions, which worked out better than I think any of us had expected. I signed on a little early, and while everyone was getting onto Zoom I would talk about my nature walks and what I'd seen, just to connect with everyone and make the class a little more humane. Maddie, my labradoodle, often joined me in my study for the lectures; she sometimes made an appearance on Zoom, too, much to my students' amusement.

I also taught a first-year seminar in the fall with a small group of eight students. The seminars are very informal and center around a shared interest of a faculty member and the students; one of the goals of the seminars is to give new undergraduates a chance to get to know a professor in a smaller setting than the large, required science classes. I taught first-year seminars for most of my time at MIT: some years we talked about how trees work and other years about how birds work. In past years, for a project, I had the students make posters or demonstrations which we donated to Mass Audubon's Boston Nature Center. (Jeannie had encouraged me to include a community service project in the class.) During the pandemic, with the students cooped up at home, I decided instead to have them keep a nature journal. Their homework was to go out for walks and take photos of what they saw, use an app to identify birds, animals, and plants, and then report back to the group at our weekly meetings. Some of the students told me this was the only time they got out of the house all week; they enjoyed sharing what they'd seen with other students in the class.

By the fall of 2021 I felt that I could no longer be "all in" at MIT, so I decided to retire. It took some time to get back to working on this book, but eventually I got dug in and I loved it once again. As my grieving for Jeannie has eased, I look back on my relationship with her with immense gratitude for all that we had and for her love for me. And as I look forward, there is more joy to be found in nature, in friendships, and, if I'm lucky, in love.

REFERENCES

Birkhead T (2016). *The Most Perfect Thing: Inside (and Outside) a Bird's Egg*. Bloomsbury.

Hanson T (2011). *Feathers: The Evolution of a Natural Miracle*. Basic Books.

Weidensaul S (1999). *Living on the Wind: Across the Hemisphere with Migratory Birds*. North Point Press.

ACKNOWLEDGMENTS

This book benefited in many ways from many people. Looking back, every time I needed something—advice about the science of birds or writing, a photograph, or just plain old encouragement—people stepped up to help. I am deeply grateful to all.

The ornithologists and biologists that I met with were wonderfully generous with their time and enthusiastic support of this project. Rick Prum of Yale University gleefully shared his love of feathers and gave permission to use the micrographs of the eastern bluebird feathers in chapter 1. Cassie Stoddard hosted visits to Princeton University, commented on an early draft of the egg chapter, and gave permission to reproduce a modified version of the remarkable plot of egg shape that she and her colleagues published; it's been a real pleasure getting to know Cassie and learn about her amazing work. I have enjoyed many delightful visits with Sharon Swartz of Brown University to discuss birds and bats; Sharon also reviewed an early draft of the flight chapters. I'm also grateful to Andy Biewener of Harvard University who read a later draft of the flight chapters and made helpful comments. David Winkler of the Cornell University Lab of Ornithology read a chapter on migration; I'm grateful to him for this, even though that chapter didn't make it into the final version of the book. David also gave permission to reproduce his photo of an egg yolk in the egg chapter. Mya Thompson, also of the Lab of Ornithology, hosted my visits to the Lab in the early stages of writing this book, and then again later, when I gave a talk on feathers. Their knowledge and advice made this a better book; any errors that remain are mine alone.

One of the challenges of this project has been writing for a general rather than an academic audience. As I researched the book and began writing, I focused on the science behind how birds work, resulting in a draft that was too much like a textbook. Knowing that this wasn't what I was aiming for, I thought that I needed an editor to help with the writing. I contacted Tom Levenson, of MIT's Program in Science Writing, with whom I had already discussed the project, who suggested Toby Lester. Toby's thoughtful questions, comments, and edits provided a master class in science writing. I'm deeply grateful for his advice and enthusiastic encouragement.

The feedback I got on drafts of the book from friends (thank you Alison Curtis, Jeannie Hess, Caitlin Stier, Susan Brand, and Andrea Fleck Clardy) and at talks I gave for birders (at the Massachusetts Audubon Society, Hog Island Camp of the National Audubon Society, the Arnold Arboretum of Harvard University, and at local bird clubs) was invaluable. Producing my monthly *Nature Notes* blog during the past few years has improved my writing; thanks to the many friends and colleagues who commented on my posts.

I had many helpful discussions about the title for the book with Susan Brand, Andrea Fleck Clardy, Jon Clardy, Terri Favro, Toby Lester, Nicholas Steenhaut, and Cathy Welsh. Andrew Kinney, my editor at the MIT Press, and I distilled their thoughts and our own into the final title.

I could not have brought this project to fruition without the steadfast support of MIT. My colleagues' and students' enthusiasm and support, especially that of my department heads, Chris Schuh, Jeff Grossman, Caroline Ross, and Polina Anikeeva, over the last nine years, has meant a great deal to me. I had the great pleasure of working with the wonderful Isaac Cabrera, then an undergraduate student in my department, who took the beautiful scanning electron micrograph photographs of feathers in figures 1.2, 1.21, 1.25, 2.12, and 2.14 as well as the photographs in figure 6.6. Isaac Cabrera and Geetha Berera took the photographs of water droplets on various materials (figures 2.3 and 2.7). The scanning electron microscope images of iris and cattail leaves (figure 3.14) were taken by Don Galler. The students in my first-year seminar How Birds Work were a delight and helped me think about how to explain the ideas in this book. In typical MIT fashion, two of the students in the seminar, Juliana Chew and Sophia Mittman, 3D-printed a single barb of a feather (using a synchrotron scan provided by Rick Prum). Finally, this book would not have been possible without financial support from MIT through the Matoula S. Salapatas Professorship and the MacVicar Faculty Fellows program, for which I am deeply grateful.

When I first approached Amy Brand, the Executive Director of the MIT Press, about my interest in explaining how birds work from an engineering perspective, I was debating what to do with the idea: whether to write a book, make videos, or write a series of articles. Within half an hour of describing my project, Amy declared "We'd publish that book." And that was it—I moved ahead with the book project. My editors at the Press, Bob Prior, and now, following Bob's retirement, Andrew Kinney, saw the potential of my idea for this book and worked with me to make it happen. Pamela Quick answered all my detailed questions about permissions. Matthew Abbate copyedited the book, his eagle eyes catching glitches that I had overlooked. The design of the book was overseen by Yasuyo Iguchi. Thank you to everyone at the MIT Press who helped bring this book to fruition. I'm also grateful for the comments of two anonymous reviewers.

Libraries are essential to any effort like this. I count myself lucky to have easy access to the extensive collections of the MIT and Harvard University libraries. The staff of MIT Libraries as well as Mary Sears at the Ernst Mayr Library of the Museum of Comparative Zoology at Harvard were especially helpful in tracking down references for me.

This book benefits from the many photographs and illustrations that help explain the ideas in the text. The figures were expertly prepared by Beth Beighlie, assisted by Grace Park. The superb drawings of birds were done by Erica Beade, whom I first met as a student in her drawing classes at the Harvard Museum of Natural History. Beth, Grace, and Erica: it was a joy to see the figures take their final form and the book take shape. Thank you.

I am especially grateful to be able to include photographs of feather, bone, and egg specimens from the remarkable ornithology collection at Harvard University's Museum of Comparative Zoology. Jeremiah Trimble and Kate Eldridge of the Ornithology Department at the MCZ were always eager to help me with loans of feather specimens for scanning electron microscopy. Jeremiah photographed the egg and bone specimens from the MCZ. Jeremiah also kindly checked the bird identifications in the figures throughout the book. Mark Omura of the Mammalogy Department at the MCZ photographed the white-tailed deer metatarsus that appears in the bone chapter. Permission to reproduce the images was given by The President and Fellows of Harvard College. The University of Colorado Museum of

Natural History provided the scanning electron micrograph of an eggshell; Jacob Veldhuizen and Karen Chin there patiently searched for the image I needed. The American Museum of Natural History gave permission to reproduce the image of the hornbill bone; Paul Sweet was helpful in arranging this.

Andy Biewener of Harvard University's Concord Field Station provided the pigeon bone specimens for micro-computed tomography imaging, which was skillfully done by his doctoral student at the time, L Fahn-Lai. The Massachusetts Audubon Society provided the barn owl feather we used for the scanning electron microscope images in figure 1.21. Silke Nebel and Dan Jackson looked through their 20-year-old files of photographs of *Calidris* sandpiper bills to find images that I could use in the bill chapter; I'm grateful that they were willing to take the time to do this.

Most of the photographs of birds in the book are from the stock photography company Alamy; Lindsay Passet at Alamy was always ready to help, making the process easy. I also appreciate being able to include the wonderful photographs of Bob Mayer (great horned owls at Forest Hills Cemetery), Norm Smith (snowy owl at Boston's Logan Airport), Tim Laman (Wahnes' parotia), and Rob Palmer (peregrine falcons). Brookline Frames provided several mat boards, cut to the same size and free of charge, for figure 3.8.

Myrah Bridwell of Cornell University's Lab of Ornithology assisted with permissions to reproduce images from *The Cornell Lab of Ornithology Handbook of Bird Biology*. I'm also grateful to the following publishers for permission to reproduce figures: American Association for the Advancement of Science, The Company of Biologists, Elsevier, Institute of Physics, The National Academy of Sciences, Oxford University Press, The Royal Society, Springer Nature, and Wiley. The availability of open access journals is most appreciated.

The beautiful website for the book, birdsupclose.org, was produced by Richard Clinch: thank you for your artful design.

I was so lucky to have had Jeannie in my life—when we went for walks, we would tell each other we needed to pinch ourselves, our life together was so great. When she was ill I used to say to her that we won the love lottery. I don't know what I did to deserve Jeannie, but I will always be grateful to her for all her love and support.

To the many friends and family who visited, called, sent cards, videos, and texts (thank you, Di, for texting me a lovely photo every morning since her diagnosis), baked bread every week (thank you, Mary), became our personal IT advisor (thank you, Beth), gave Jeannie acupuncture at home (thank you, Karen), brought meals, and did errands for us during Jeannie's illness: there are no words to express our gratitude. Dale Archer and Matilda Mutty, our two main home health aides, are saints—thank you, I don't know how we would have managed without you. Your caring eased Jeannie's last year and helped get me through the most difficult time of my life.

Finally, a special thank you to Susan Brand, who pushed me to get back to writing this book. Thank you to everyone who helped me get this book over the finish line.

A portion of the author's royalties from the sale of this book are being donated to the Massachusetts Audubon Society and the Cornell Lab of Ornithology.

NOTES

PREFACE

1. Government of Canada website: https://natural-resources.canada.ca/our-natural-resources/energy-sources-distribution/electricity-infrastructure/about-electricity/7359.
2. For those who would like to know more about Robert Hooke and his book *Micrographia*, see the video I made with MITx, at: https://ocw.mit.edu/courses/3-054-cellular-solids-structure-properties-and-applications-spring-2015/pages/related-videos/.
3. A video on how woodpeckers avoid brain injury, *Built to Peck*, with eight two- to three-minute segments, based on my talks, is available at: https://www.youtube.com/playlist?list=PLTz-v-F773kXzGHaX1-zV6z0MPCxV8GJM.

CHAPTER 5

1. https://www.audubon.org/news/what-dcides-clutch-size; http://www.discoverwildlife.com/animals/birds/do-birds-count-eggs-their-nests;https://onlinelibrary.wiley.com/doi/pdf/10.1111/ibi.12328.
2. https://www.npr.org/2016/05/12/477827156/npr-live-lab-how-strong-are-eggs-we-walked-on-them-to-find-out.

CHAPTER 6

1. See http://ci.moraine.oh.us/the-pinnacles/. A plaque near the site reads:

   The Pinnacles, formerly located just west of here, consisted of a gorge with unique large boulders created during the ice age on its slopes. . . . The updraft created by the terrain attracted soaring birds. From 1897 to 1899 the Wright Brothers biked here and observed buzzards (turkey vultures) soaring gracefully above the river valley. In the summer of 1899 the Wrights developed their wing warping theory while watching the birds at Pinnacle Hill twist the tips of their wings as they soared into the wind.

2. Headline, "It's so hot in Phoenix that airplanes can't fly," *Washington Post*, June 21, 2017, Amy B. Wang.

CHAPTER 7

1. Galapagos penguin speed: https://galapagosconservation.org.uk/wildlife/galapagos-penguin/; https://www.smithsonianmag.com/science-nature/14-fun-facts-about-penguins-41774295/; Watanuki et al. (2006).

INDEX